March 10, 1999

To Sam,

With many thanks for all your patience and persistence in bringing my tennis skills to a higher level. This will make a mixer out of you!

All the best,

Mari-Jo

ON THE HILL

Reflections of a Miner

Profits from the sale of this book go to The Dominican Order of The Sisters of Charity in La Oroya, Peru, to aid in their work with the oppressed and victims of terrorism by El Sendero Luminoso (The Shining Path).

To all miners on a hill.

Brian Richard Woolfe, 1926 – 1988.

ON THE HILL

Reflections of a Miner

By

BRIAN R. WOOLFE

WESTERNLORE PRESS . . . 1993 . . . TUCSON, ARIZONA

Library of Congress Catalog Number 93-060195
ISBN 0-87026-084-7

PRINTED IN THE UNITED STATES OF AMERICA BY WESTERNLORE PRESS

TABLE OF CONTENTS

Foreword

The Cerro de Pasco Corporation, named for the town where its first mine was located, ran its mining enterprise in Peru from 1902 until it was nationalized in 1974.

Mining men and their families came from all over the world to work for Cerro, usually with the idea of working one three-year contract to make some money and then return back home. More often than not, after a long paid leave of three months following their contract, they would eagerly return to the Andes of Peru, to what had become their real home, and remain for many years. At least one Cerro family was "on the hill" for three generations. These "gringos" were true expatriots, far happier living in Peru than in their own countries of origin.

In a mining career that spanned four decades and five continents, Brian worked for almost twenty years on the hill in Peru for "C. de P." His reason for leaving in 1970 was to become Project Manager, then first General Manager when production started, of Cerro's Rio Blanco Mine in Chile.

Born in Cornwall, England, Brian grew up in India where his father was an officer in the British Army. He served as a midshipman in the Royal Indian Navy during World War II. On his return, he entered Camborne School of Mines in England, working during vacations in mines in France, graduating as a Mining Engineer in 1949. His first full-time job was in Western Canada, his last position was in Arizona as President and C.E.O. of Magma Copper Company.

He enjoyed writing of his experiences in his free moments,

and had in mind something that would be an anatomy of a mining operation (specifically Cerro's in Peru in the '50s and '60's unless otherwise noted) written in a way that a person with no mining or engineering background could grasp.

Most of what he planned to write is here with the exception of his thoughts on present day mining problems, and some stories set in the outside camps, the main camp being in La Oroya where the smelter and head office were, and about the annual rodeo held at the Casaracra Station of the company's sheep ranch, in itself an exceptional operation.

Brian had a great respect for his profession, whenever asked what he did, he'd reply simply, "I'm a miner." Shortly before we were married he announced to me, "I have something I must tell you, I have a mistress. It's not another woman, you never have to worry about that, it's the mine. I'd never get up at 3:00 a.m. for any woman, but I will for the mine." The following pages should make that obvious.

Mari-Jo Woolfe

Chapter One
Getting Started

The first mine I worked in South America was located over 14,000 feet above sea level at Yauricocha in the Peruvian Andes. This may look like a misprint, but it's not. There are quite a few mines as high or higher in Peru, and the highest working level I ever had to look after was a few feet over 16,000 feet.

In the Central Peruvian Andes, the landscape at these altitudes is barren of all trees and mostly consists of rock, scree slopes, lichens and tough stunted grass. From May to November it is usually bright and sunny with very cold nights, from November to May there is frequent rain and snow. As this is the southern hemisphere, it is generally warmer during the rainy season than during the sunny months. Snow usually doesn't stay long, and eight inches cover on the ground is about as much as one will ever see.

The mine was small, but very rich, in fact at that time the ore was sent directly to a smelter to be converted into metal, and did not have to be concentrated first at the mine site.

The mine crew consisted of about fifteen gringo staff, mostly Americans and Canadians with some Europeans, and national Peruvian engineers and about five hundred Indians. The Indians were short, barrel-chested, and quite able to do physical labor at these altitudes, something which no gringo ever managed to do for any length of time.

My first attempt to show an Indian miner how to drive six-inch nails with the back of an axe, something I had learned to do well in Canada, ended ignominiously. I was doing fine, driving each nail with three clouts, when I started running out of breath and this red haze with black dancing spots appeared before my eyes. After some ten nails had been well and truly driven home, efficiency dropped off remarkably, and finally after a few last futile flaps of the axe in the general direction of the nail I gave up the demonstration. The miner was highly amused.

Altitude sickness, or "soroche" as it is locally called, afflicts just about everyone when they first arrive in one of these high camps. The effects and intensity though seem to vary from person to person, and even in the same person can change from time to time. Some people never adjust and have to leave with a medical discharge. One of the effects of the scarcity of oxygen is that the red corpuscles increase in number and occasionally this process continues until the blood becomes so thick that the heart is overburdened and one has to go to a lower altitude.

The most peculiar case of soroche that I recall was that of a young Rhodesian supervisor. On arrival at the mine the Rhodesian was young and fit and keen, and with some experience on the Zambian copper belt. Just the type we were looking for. However, after a few weeks his enthusiasm seemed to wane and we noticed that he would eat his lunch quickly, then sack out in the nearest chair until it was time to go back to work. Then we heard that he also slept in the station wagon both to and from work. It wasn't long before the mine superintendent was visited by his tearful wife. Between sobs, she explained that her husband went straight to bed after returning from work, and had started sleeping through the weekends from Saturday until Monday. He was given a medical discharge as we were afraid he might fall asleep while walking around his beat in the mine and injure himself!

Another affliction that a newcomer frequently encountered was stomach trouble. As any tourist will testify, this is not a peculiarity of the high Andes, and is common to most countries as verified by its various names: "Montezuma's Revenge," "The Inca Two-Step," "Delhi Belly," etc. It was usually of short duration, and once one's stomach and digestive tract had become accustomed to the new food, bugs, and water it ceased to be an annoyance.

This mine had ore bodies, or "mantos," which were irregular fairly large mineralized ore masses containing copper, gold and silver values. Also in some areas there were zinc and lead values along with the copper. As copper is more valuable than lead or zinc, these latter areas were not mined until later, after a concentrator had been built. The mine was being worked from the top down, and many of the ore bodies went right through to the surface. This procedure is usual in underground metal mining as it enables one to get into production quickly, and the ore from the upper levels then helps to pay for the development of the lower levels.

The ore was hoisted up a shaft that was about 700 feet deep in buckets called "skips." The shaft also had "cages" which are the conveyances used for hoisting and lowering men and materials. This was a pretty shallow shaft as shafts go, in South Africa they have very deep mines going down to 10,000 feet below the surface, and some shafts are 5,000 to 6,000 feet deep. When going to these extreme depths it has been found more economic to use three or even four shafts to get to the bottom levels, and 5,000 to 6,000 feet is about the limit for one vertical shaft section.

The hoist cables do break occasionally, although this is now a very rare occurrence with modern day maintenance and safety regulations. I once had the misfortune to temporarily take over a mine while the superintendent was on vacation, and on the third day of my temporary promotion a cable broke.

Luckily, the cable broke close to the skip, so although the skip was loaded with eight tons of ore, it fell 2,000 feet without doing much damage because there was no great length of cable

The Amazon
Ucayali River
Ucayali River
PUERTO BERMUDEZ
TINGO MARIA
6,500
Huallaga River
Perené River
LA MERCED
HUANUCO
AMBO
HUAGARUNCHA
17,850
TARMA
Marañon River
GOYLLARISQUISGA
13,700
CERRO DE PASCO
14,200
L. JUNIN
JUNIN
CERRO
RAURA
OROYA
12,200
Mantaro River
MALPASO
DAM
PACHACHACA
MAHR
HUARON
15,090
RAURA
MINA RAGRA
MOROCOCHA
14,850
SAN
CRISTOBA
15,500
TICLIO
OYON
CASAPALCA
13,730
CHOSICA
2,800
Rimac River
LIMA
CALLAO
N

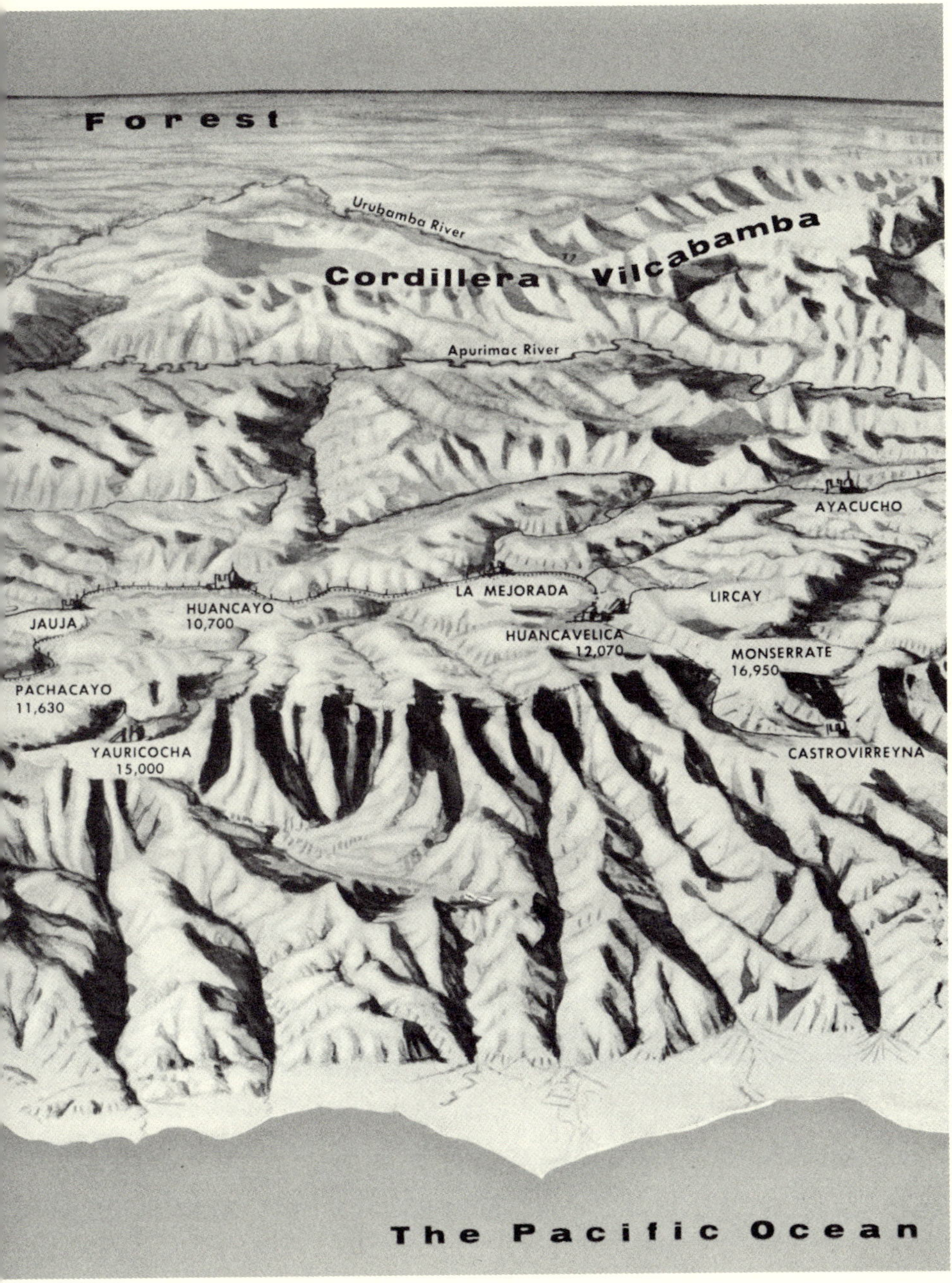

Forest
Urubamba River
Cordillera Vilcabamba
Apurimac River
AYACUCHO
LA MEJORADA
LIRCAY
JAUJA
HUANCAYO
10,700
HUANCAVELICA
12,070
MONSERRATE
16,950
PACHACAYO
11,630
YAURICOCHA
15,000
CASTROVIRREYNA
The Pacific Ocean

whipping backwards and forwards tearing out the shaft timber behind the falling skip. At the bottom of the shaft it was a different story, the scene resembled a bowled out match factory. Fortunately, no one was hurt although the two men who had been loading the skips on one of the bottom levels were both knocked flat by the air blast when the skip rocketed past.

As the mine was near the top of a mountain with fairly steep relief, some of the upper levels were connected direct to the surface by tunnels. These are called "adits." The levels were about 180 feet apart and between them were the "stopes" or working places which produced the ore. The crews on the levels collected the ore from the stopes, in this case with electric locomotives pulling three-ton cars, and transported it to the ore pockets at the shaft from whence it was hoisted to the surface and dumped into bins.

One cannot, except in special cases, go on removing large quantities of ore and leaving immense voids without asking for trouble. Consequently, waste rock or gravel is introduced from the surface to fill these voids. The material is called "fill" and this operation is called "filling."

One good precept to remember is that one can usually handle bad ground over one's head by various types of support, but that any movement underfoot, in or below the floor, is extremely dangerous and very difficult to handle. About the only method is to bore holes into the suspect area and then pump material in suspension (grout) to fill all the voids and consolidate the ground. In many cases this takes thousands of sacks of cement, and is a long and costly process.

The stopes in this particular mine were about 15 ft. x 60 ft. long. As the ore was broken out, the ground was supported by timber sets. After two horizontal slices had been removed, the bottom level of sets would be filled with waste, leaving open two adjoining sets at each end of the stope.

Of the two open sets, one would be lined and used for dropping the ore to the level below, and the other would be equipped with ladders and a small lined compartment. This was known as the "manway" and the compartment was used to hoist timber and materials up to the stope working floor.

There are many kinds of stoping systems, of which this kind is the most expensive, but it enabled us to extract 100% of the ore. In an ore body with very high values and irregular walls, it was at the time a pretty good system.

The mine, undoubtedly, had the heaviest ground it has ever been my misfortune to encounter. Every opening had to be supported with timber sets, and a large group of men were constantly employed replacing the breaking timber. In the main haulage ways we put in 10" x 10" eucalyptus timber sets with a spacing of 10 inches and "doubled up" by putting another 10" x 10" set inside each original set. Even these doubled up sets would seldom last more than three months!

One of my first shifts, an old mining school acquaintance who had already been at this mine for a couple of years or so, took me to a stope that he remarked was "being a bit of a problem." A block of ore, calculated at some 60,000 tons, had broken free of the main ore body and was creeping down a fault line and squeezing the timbers supporting the walls of the stope. The mine crews were pouring waste rock from the surface into all the available open area, and also men were reinforcing the timber sets while other crews were cutting slots in the settling block to try to break it up into smaller and more easily handled segments. As the block slowly settled, its full weight would come on one or two timbers, which would slowly crush, then as this resistance was overcome, the full weight would come on another section of timber.

The groaning and cracking of timber could be clearly heard throughout the working area, and I found it quite the most unnerving sound I'd ever heard. It was a race against time, and as my friend remarked, the miners wouldn't stay unless we stayed with them, so we stayed. I remember vividly hearing a rather long–winded story about some red-headed girl my Irish friend had met during the war, and at the same time watching popeyed as the 8 x 8 timber over our heads gradually started breaking in half. I finally managed to get a word in edgeways, and we moved to a safer spot.

When timber creaks and groans you have time to take preventative action. When it breaks with a noise like a rifle shot, then you better leap to the nearest exit. In this particular case, the stope was saved. Once the block had been cut into smaller pieces, the timber was able to hold the smaller weights pressing on them.

Mostly, to break rock in underground mines, you drill holes in it with pneumatic machines that make a noise like a paving breaker, then fill the holes with dynamite, and blast a series of holes, or a "round," together to make your production in a stope or your advance in a tunnel. Several machines working in one confined underground area at the same time produce an ear-shattering roar, and it is no wonder many miners are hard of hearing. Legislation is only beginning to come in now in some counties with regards to using silencers on rock drills. In our mine, the ore was generally very soft, hence the large quantities of support timber, and it was easily broken using a pick-hammer, a machine that was in effect a light handheld paving breaker.

When a mine is drilling and blasting all its ore, an eight-hour shift is usually worked, the blasting done at the end of it. Then there is a four-hour break to allow the smoke and gas to clear, and following that the night shift goes to work for eight hours. They also blast at the end of their shift, and then there is another four-hour ventilation break before the day shift returns to work. As we did relatively little blasting, we did not need the time out for ventilation, so we worked two production shifts back-to-back. The third shift from midnight to eight in the morning, also called "the graveyard shift," was a supply shift that placed the large amounts of support timber on the levels and in the working places ready for the production shifts.

Brian, section foreman, with a crew at Yauricocha.

Normally, the mine worked a six-day week, with Sunday reserved for special maintenance jobs such as shaft and hoist repair.

Generally, ore from a metal mine is concentrated after it is mined, and the concentrate is sent to a smelter to make crude metal. Then the crude, or impure, metal goes to a refinery where the final product is turned out. Companies may or may not go through all these processes themselves, and in our case, as the ore was very rich in the early days, we skipped the concentrating stage and shipped direct to the central company smelter in La Oroya.

The mine lay in the Pacific watershed of the Andes, with the smelter lying in the Atlantic watershed. The Continental Divide between the two was not only high, but also contained some very precipitous territory. Consequently, we transported the ore by cableway to the relatively level altiplano on the other side of the divide. After a journey of some twelve miles by cableway, the ore completed its trip to the smelter by rail.

The cableway was a beautiful engineering job, and included two bridges which were 600 and 700 feet above their respective valley floors. The longest of the two was over 3,000 feet long, and crossing it on the swaying catwalk, with an open wire mesh deck was not recommended if you didn't like heights!

The Indian crews who ran the cableway operation and carried out any necessary repairs were completely oblivious to the height. They would stand on the steel structures that supported the cableway sheave wheels with no safety belt, no handholds, and 600 feet of nothing underneath. The catwalk and cableway supports were in turn supported by suspension cables, six on each side.

When strands started breaking in the suspension cables, we had to seize or bind the broken strands to stop the cables from unravelling, put in two additional support cables on each side, then put on wooden clamps to stop the suspension cables from banging together in the wind.

The repairmen would slide out, sitting on the six one-inch cables to do their work and then, because of the catenary of the suspension cables, they would not be able to pull themselves back to their starting point, the support towers. The only way out was to go under the suspension cables and slide down one of the support cables to the catwalk deck fifty to a hundred feet below. While doing this maneuver, no safety belt could be used. It never failed to raise the hackles on the back of my neck every time one of the crew went through this exercise.

The cableway looked impressive, but it was a bottleneck to production, and efforts to put on more buckets, and to load them more heavily brought all kinds of problems. Twice the cable broke, and as it was a mono-cable there were buckets dropped all over the countryside. We pulled miners from underground to help pick up the wreckage, and it looked like a tortoise migration as two men got under each bucket and started heading up out of the valley to the road.

The two ends of the cable were more than half a mile apart, and they had to be pulled together by tractors before splicing could begin. It was not unusual to have a bucket tangle up in a tower, then after several more buckets had joined in the pile-up, over would go the tower. Invariably, some of the mine supervisors would be called out to help with the repairs, and work would continue round the clock until the cableway was functioning. Shifts of twenty hours at stretch were often worked.

The one thing that could interrupt repairs were electrical storms. There was no sense in continuing work, perched sixty or seventy feet above the ground when the lightening started and the very air around you started to crackle. Snow was no big problem, except it made the steel slippery, and working conditions were even more unpleasant than usual.

Exploring.

Chapter Two
Exploration

The mine supervision and personnel are divided up into various departments according to their specialties and responsibilities. The Geological Department is responsible for finding the ore, assessing its value and maintaining a sufficient inventory of ore so that production planning can be carried out efficiently.

Geologists are noted for being very cautious in their estimates and in stating their opinions. Giving a long dissertation on the origins and geological controls of our ore body, which had us listening fascinated, our Chief Geologist paused, looked thoughtful for a moment, then added, "Mind you, I don't necessarily believe what I am saying!"

Another time, three of us were wandering around some old workings in an abandoned part of the mine checking out plans to see if the old-timers had left any ore which could be economically exploited at present day metal prices. As we rounded a corner, we came upon a geologist busily mapping the structures revealed along the side of the tunnel. Realizing that he had been spending some time in the area, and probably knew it better than ourselves, we asked him where a tunnel went that was not marked on our plans. His response was immediate, and for once definitive, "Oh, that goes right to the end!"

In spite of their peculiarities, geologists are extremely valuable members of the mining team, and the success of a project has frequently depended on their expertise. A frequent question is: How do they find mines in the first place? In the early days, the ore bodies that were exploited were the ones that went right to the surface or “outcropped,” and finding them was relatively easy. People noted funny colored rocks or rocks that were abnormally heavy, and when geologists or prospectors came searching for ore they heard about these ore bodies, and by carefully studying their geological environment it wasn’t long before geologists were beginning to deduce the probable whereabouts of hidden ore bodies, and over the years developed many tools to help them.

Nowadays, there cannot be many outcropping ore bodies left except in extremely remote areas of the world, so geologists and their friends, the geophysicists, have had to work more and more with techniques to discover hidden ore bodies. Whereas a geologist studies rocks and formations of the earth’s crust, the geophysicist studies the physics of the earth, such as magnetic and gravimetric fields. There is naturally a lot of overlapping of the two sciences in practice, because to find an ore body one has to know something of the other specialty.

Present day exploration teams use many different sophisticated instruments to pinpoint likely areas. This work though is usually done after the area has been mapped and the geological structures and rocks appear promising for the ore which one is looking for. As they say, “If you want elephants, you go to elephant country,” and it is the geological maps that will tell you if it is indeed the right type of country.

Among the instruments used are magnetic, gravimetric, and a system to measure electrical currents in the earth’s crust, whether existing or induced currents. The first is used for magnetic ores such as iron and some nickel ores that will cause anomalies in the earth’s magnetic field. The gravimetric instruments depend on the fact that most ores are much denser than their surrounding host rock, and will distort gravitational readings taken over them so that they don’t match surrounding readings, or thus make an anomaly. The electric methods rely on the better conductivity of say, sulfide minerals than the host rock. All these instruments have to be extremely sensitive as they are measuring minute differences.

Geophysical instruments can be mounted in planes or helicopters, and large tracts of very difficult country an be checked out rapidly. However, all they will tell you is that an anomaly is present, they won't tell the grade of the ore, or what is making the anomaly. Water-bearing faults can foul up an electrical survey in the worst way. If the mineral does not pay for the trouble of mining it, then it is not considered an ore body, so the grade is kind of important.

Photos from cameras orbiting in space, side-looking radar, and the measurement of mercury emanations into the atmosphere, all provide information for geologists and geophysicists to work with in their continual search for ore bodies. Proving new equipment and improving old equipment for exploration work is a never-ending process. To a very large extent, the future of the world's mining industry will depend on the ability to find hidden ore bodies.

Radioactive ores can be located by geiger counters and scintalometers which measure the amount of radio activity emanating from the rock. Other minerals are invisible to the naked eye in their surrounding rock, but under ultraviolet light glow with different colors and intensities. Scheelite, one of the tungsten ores, is one of them.

The search for ore also extends below the surface of the seas, where areas of the sea bottom rich in manganese nodules containing copper and other metal values have been located in the Pacific, Atlantic and Indian Oceans. In the depths of the Red Sea, and in the central Atlantic, areas have been located where it seems ore bodies are being formed by mineralized solutions coming from the earth's magma.

Geochemical methods are also used to measure the amount of different metals present in soil or in plants. Plants, some more than others, have a tendency to concentrate metals in themselves over a period of time. These measurements all have to be made in parts per million, and when an area is discovered where the values are higher than normal, it is investigated more closely.

Once a geologist has decided on a likely area, he needs to look at samples, and to get these samples he uses a diamond drill. This is a machine that rapidly rotates a hollow pipe-like drill rod with a bit on the end that contains industrial diamonds or "borts" set into the metal. The bit is also hollow like a piece of pipe, and the diamonds are set just in the rim and edges of that rim. As the bit rotates, pressure is applied and a cylindrical section of rock or core is cut free. This core gradually fills the core barrel as the bit advances through the rock. When the drill has advanced sufficiently to fill the core barrel, then the driller pulls his string of rods and removes the sample. He then returns the string to the hole and adds a new section of drill rod in order to advance another ten feet or whatever the capacity of the core barrel may be.

Water is pumped down the hole while drilling in order to cool the bit and also to flush out the drill cuttings. The cuttings or sludge is saved, as it is also an indicator of the ground that the bit has been passing through. In very bad ground it might be the only record you have left, as you may get no core at all. In fact, to get between 90 and 100% recovery is exceptional, and often one has to be satisfied with a lot less. New systems and equipment are improving the efficiency of drills all the time, and the wire-line drills enable you to extract your core without pulling all the rods to the surface.

Diamond drill holes have been drilled to well over 5,000 feet, although it is not common to go to such depths. The Orange Free State gold field in South Africa is an example of successful mines being found by a deep-drilling program, 4,000 to 6,000 feet, based on geological deduction and careful mapping.

Once a promising ore body has been outlined with what looks like high enough values and of sufficient tonnage to sustain a mining operation, the miners and metallurgists come into the picture. They are the first to decide on the probable type of mining method to be used, and then to calculate its cost per ton. Very often in mining the ore will become diluted and the value per ton of ore extracted will not be as high as the geologist's original estimate. Then the metallurgists run checks on the samples to decide what method of concentrating would best suit the ore, what the cost per ton would be and what would be the likely recovery. Here again some metal value will be lost.

Finally, if everything still looks good, a feasibility study is made which will include financing, marketing, services (such as power and water supply, communications, roads, railways, etc.) and taxes. In the case of overseas properties, an appraisal of the political situation and the risk of future nationalization, and/or special mining law and government conditions must also be evaluated.

In maybe one out of a thousand cases, depending on the area you are looking at, your original anomaly may turn out to be a mine. Even so, a sudden drop in metal prices might make your property marginal or even put it out of business.

Meanwhile, the geologist is still gathering every possible scrap of information possible about the ore body. Once he can establish the reasons for the ore being where it is, he has the vital clues that could lead him to extensions of the known ore body, or perhaps to other completely new deposits in the surrounding district.

Ore bodies are wasting assets, new ore does not grow as mining proceeds, consequently once mining starts the geologist must constantly be looking for new reserves to replace the ore that has been mined. In a big operation, such as International Nickel at Sudbury in Canada, this can amount to a million tons of new reserves having to be found, sampled and proved to exist beyond reasonable doubt every month.

Once mining starts, the geologist samples all the working

places on a routine basis, and also maps all the geological structures as they are revealed, day by day in the fresh ground being exposed. He also will continue to diamond drill as necessary to gather more information. All this data is interpolated by the Geological Department after it has been put on plans and sections, and is utilized by the Mining and Engineering Departments in their short and long range production planning.

In our case in Peru, the geologist was also responsible for checking the surrounding district for new prospects, old mines and small working mines. This was both to improve his overall knowledge of the district and its geological characteristics, and also to keep an eye open to possibilities for new mines. In the high Andes this was a relatively simple task, as much of the geology was easily seen in the exposed rock formations, and travel was not so difficult as, say, in a forest. However, one could ride for hours with an Indian guide to finally discover that the "bonanza" he has been telling you about has shrunk to a few veinlets of useless pyrite, also known as "fool's gold."

In western Canada one could have some unpleasant surprises. Two of my friends, complete with heavily loaded backpacks full of samples, encountered a grizzly bear on the way back to camp. They very smartly headed for the nearest tree and climbed it as fast as possible, this is because black bears climb trees, but, for some reason, grizzlies don't. The interesting part was that the last guy to arrive at the base of the tree was the first to the top. Being the closest to the bear, he was the most scared, and had climbed right over his partner!

Also in British Columbia, a geologist friend and a local prospector were investigating some old workings in the Selkirk mountains. As these tunnels were made by hand drilling at the turn of the century, they were both low and narrow, and the two Canadians, one of whom was an ex-professional wrestler, were finding it rather a tight fit as they tramped along with the aid of a flickering carbide lamp. After they had advanced several score feet, at the very edge of visibility cast by their lamp, they saw something moving. As this was springtime, and bears were starting to wake from their winter hibernation, the men turned around and went out of there like a shot. In spite of the small dimensions of the tunnel, the one who had been leading the way managed to overtake and pass his companion, and be the first out to the fresh air. It turned out afterwards that the suspected bear had been a porcupine, who quite often go into old workings.

In the Andes, the villagers in the area surrounding the mine were usually very helpful and hospitable to any visiting geologist. They realized that a new mine in their locality would not only bring work and wealth, but also roads and schools and medical facilities. One engineer had been checking a possible road route to a new mine during the course of several months. This particular time he approached a village, which he had visited before, while he was suffering badly from dysentery. Just as he started to enter the village he felt another painful attack. Jumping off his horse, he rushed behind the church and settled behind a gravestone.

Unbeknownst to him, his approach had been noted by a small boy who had informed the mayor that the gringo was on his way to visit. On hearing this, the mayor collected the other village notables, and with the usual crowd of curious onlookers, went to welcome his guest. Trying to acknowledge a civic welcome crouched behind a tombstone, racked with stomach cramps and with his trousers around his ankles was an experience the engineer would not easily forget.

After several days by horseback on one trip, we arrived at a campsite in the evening alongside some thermal springs. The Incas had built a series of baths with large blocks of stone, the water being successively hotter in each rock bath as you got nearer to the source of hot water. Once we had pitched camp, we joyfully shucked our clothes, grabbed soap and towel, and piled into the hottest bath we could stand to ease our aches and pains and to remove the dust of the trail. We were halfway through our ablutions when the local village notables turned up accompanied by numerous onlookers.

We quickly sank down into the baths until the water was up to our necks to accept the welcome. The speech droned on and on while we were gradually being parboiled. Finally, bright pink and gasping, we cast modesty to the winds, jumped out of the bath and rushed for our towels. The mayor never blinked an eye, nor faltered in his speech; the perfect gentleman. However, high-pitched giggles from behind him proved that not all the crowd were so courteous and phlegmatic.

If you really got off the beaten track, then things were a little different. It would be unlikely that the people you met spoke any Spanish for a start. Then many of them believed the story, incredible though it may seem, that the gringos captured Indians and boiled them down to make grease which they used to run their machines. The first thing when you approached, they would try and run away, then when they realized they couldn't escape men on horseback, the women would squat down and start throwing dirt up under their skirts, presumably to discourage any thoughts of rape. Information from these Indians would usually be unreliable, as they would say anything just to get rid of you.

Chapter Three

Mining Department Relationships

The Engineering Department of a mining company is responsible for all the surveying, i.e., making sure that the tunnels go where they are supposed to and connect where they are supposed to. As part of this work they prepare all the maps of the working places, the claim maps that show the mine's mining concessions, and lay out the exploration drill holes for the geologists. They also work with the geologist and miners on the preparation of production plans.

The Mining Department is the biggest department and has the responsibility for producing ore of the pre-determined value and tonnage to meet the production targets. Usually the mine is divided up into conveniently sized sections and a junior supervisor is made section foreman with one or two shift bosses and fifty to one hundred or more miners to do the work and make the production. The section foremen report to a mine captain or assistant mine captain depending on the size of the mine.

I believe the term "mine captain" comes from the old Cornish miners or "Cousin Jacks" as they were known. Certainly as late as 1946, when they talked about the depths of their shafts in Cornwall they used "fathoms" instead of feet or meters.

At the beginning of each shift the section foremen, after reporting to the mine captain and checking the log book to see what comments the previous shift has on the various working places, line up their shift bosses and work crews. They then go underground and visit the men in the working places, inspecting and directing the work and taking notes for the mine captain and next shift.

The miners never worked alone, and the smallest work group would be two men. The idea was basically one of safety, if one man was injured the other would go for help. Sometimes if they were unlucky they both might get injured or killed at the same time, luckily this rarely happened. The men for the most part worked hard and willingly and were deeply interested in their work. As they were alone in their working places most of the shift, with only brief visits from their supervisors, they were paid a bonus system, over and above their base wage, depending on the number of tons they produced or the amount of timber they installed. Weekly, the production results of each stope were posted on a blackboard, and the men would eagerly check to see how they were doing.

The section foremen had the toughest jobs amongst the gringos in Peru, not only were they climbing up and down into the working places most of the shift, and this at altitudes from 12,000 to 16,000 feet above sea level, but they were doing it in atmospheres too often tainted with rock dust, gas and smoke from blasting. On top of this, the temperatures could vary from freezing point to around 100°F, often with high humidity to boot. In some mines the massive sulfide ores would give off considerable amounts of sulfur dioxide which would make breathing most difficult. Wet working places were not at all uncommon, and one could easily get soaked to the skin early in the shift and then have to go to some freezing cold area. Although it was not quite as bad as Canada, where at temperatures of –40°F and –50°F your clothes would freeze solid in a matter of minutes.

The water and sulfide ore very often produced a sulfuric acid of around 2.5 pH which was extremely painful if it got in your eyes and would burn holes through your clothes in no time flat. In bad acid areas the only answer was to wear acid–proof coveralls.

The section foreman would also be subject to emergency callouts at any time of the day or night if there was an accident or problem in the mine. No wonder the men who did this work

Aerial view of Cerro de Pasco mining complex and town.

considered themselves better than the rest of the staff, and also were inclined to be wild and wooly in their off duty hours. They liked nothing better than to take some unsuspecting accountant or storekeeper underground on a visit. If they were going to a hot area the visitor would be told to dress warmly, if going to a freezing cold wet area he would be told the opposite. All the dirtiest, gasiest and most miserable working places would be chosen for the trip, and every one would involve a lot of ladder climbing.

The accountants sometimes got their revenge. One mining engineer signed for a large open pit shovel costing some $400,000, and inadvertently used his personal account number instead of his operational code number. At the end of the month there was no pay check for him as he was hundreds of thousands of dollars in the hole. To add insult to injury the accountant dutifully and delightedly followed company rules and cut off all credit in the club and company store for the engineer who made the mistake. It took several weeks to sort it all out, and the accountant, who had been the victim of several trips underground with the man concerned, dragged his feet all the way.

Usually a close relationship grew between the section foreman and his crews, and a fierce loyalty to each other developed. This led to healthy rivalry between the sections, but sometimes this rivalry went too far. In one case, two sections had the same level, one east of the main shaft, and the other west of the main shaft. As there was a shortage of mine cars on the level at that time, a constant stealing of each other's cars took place. This supposedly reached its climax when one of the section bosses cornered most of the cars in his area, and then tore up the connecting track between the two sections. However, the other boss was a man of resource. He stayed late the next night with a picked crew, and by means of a winch, he stole back all the cars he needed plus some extra, and then blasted down the roof of the connection between the two sections. At this point the mine captain stepped in.

Another time a new supervisor came to one of the upper sections, he was new to that mine, but had considerable experience in other company mines. Shortly after he took over, the production of the section dropped off quite sharply, and the mine captain started to give him a bad time. About the same time the production of the section below started to increase, and the supervisor in charge was continually being pointed out by the mine captain as a wonderful example of a good miner and good organizer to the other section bosses.

Several months passed, and the upper level supervisor was transferred under a cloud as not being "production oriented." A section boss who had previously run this area was then transferred back to it and immediately, in fact overnight, production returned to somewhat better than it had been in the old days. Furthermore, the production of the lower section surprisingly dropped off sharply.

Smelling a rat, the mine captain decided to personally investigate his paragon of virtue on the lower level. It soon came to light that the newcomer to the upper level had inadvertently directed his locomotive crews to drop part of the production down the wrong ore pass. Consequently, this tonnage never got counted as his production. Our hero on the lower level, suddenly discovering this manna from heaven appearing in a hitherto unused ore pass, had kept his mouth shut and added it to his own production. His extra days off and other small perquisites came to an abrupt end. The victim never did find out what had happened to him.

The Indian miners were a happy-go-lucky lot with a good appreciation of dirty jokes, usually the dirtier the better. Mostly they were bilingual, in that they spoke both Spanish and Quechua in their local dialect. When the Incas conquered a new territory it was their habit to break up the conquered tribes and distribute them throughout their more settled areas. The new border territory was then settled with reliable villagers. This had the result of making the Quechua language pretty universal throughout Peru. However, there were certain remote villages which never did get conquered, but remained as isolated islands in the Inca empire. One of these villages was not too far from the mine, about three days journey by horse and foot, and these people only spoke Cauci, no Quechua and no Spanish.

Fortunately, we never had too many of these strangers present as communication was too difficult. They would turn up for work with a pouch of corn kernels at their belt. Each day they would throw away one kernel until eventually the pouch was empty. They would then quit as they knew they had completed the length of time they intended being away from home. Because of this habit the other Indians called them "Camcha Luncas" (loosely translated meaning crazy corn people), which was also used as a term of derision for anyone they thought was particularly dumb.

We had one curious case of the miner without a language. One day while visiting the working places accompanied by my shift boss, I enquired how a certain miner was making out in his working place. This shift boss told me that there were problems because his helper couldn't understand any orders. Thinking that the man only spoke Quechua or perhaps Cauci, I asked why he hadn't been transferred to work with a miner who could speak Quechua. The shift boss replied that it appeared the new man only spoke English. Somewhat taken back by this, and thinking that perhaps we had a destitute Englishman or American on the payroll, I changed course and headed directly for the working place to which the stranger had been sent.

Eventually we arrived and there was no doubt that the new man was a true Sierra Indian, so I addressed him in Spanish to which he replied in very halting English. It took awhile to discover his story as his English was so limited and poor, but piecing it together bit by bit he came out with the following tale.

At the age of about eight he had been taken from the Sierra to California by an American missionary family, and had lived with them for several years as a servant. In his teens he had left the family, he couldn't or wouldn't give a reason for this, and had gone to work as a steward on a coastal steamer.

Finally, the immigration authorities had caught up with him, and as he had no immigration papers, they had unceremoniously shipped him back to Peru virtually penniless. During his time in the States he'd forgotten all his Spanish, and at his social level in Peru found no one who could speak English. The curious thing was, that while he was gradually forgetting his English through lack of use, he didn't seem to be picking up any Spanish. He didn't stay with us very long, and later I heard that he had worked at another mine for a short while and was still having the same problem. I often wondered what became of him.

Although the Indians were happy-go-lucky, they were like most miners around the world, superstitious. They did not like women or priests going underground. However, it was alright for a priest to bless the mine once a year on the surface, where most mines had a small grotto. I must admit, although I'm not superstitious, I don't like women going underground either. Twice in Canada after women had been underground we had fatal accidents. I'm not superstitious, just careful.

The Indian miners, along with miners in the old mining areas of Cornwall, Germany, Sweden and Russia, believed that little men, whom they called "mukis," lived in the old abandoned workings of the mines. If a tool was stolen or something misplaced, they would shrug their shoulders and attribute it to the mukis. In certain areas of the mine there would be places where they would leave gifts of food, coca leaves or cigarettes for the little people. A friend of mine who used to help himself to these cigarettes whenever he ran short underground, was killed by a cave-in on his first contract in Peru.

The coca leaves mentioned above are leaves from the coca plant from which cocaine is derived. Many of the Sierra Indians would stick a wad of leaves and a little lime in their mouths and then slowly masticate it. It was a mild soporific and would deaden their feelings so that they could endure hunger and cold more easily. Luckily, by dint of a lot of propaganda, its usage in the mines has largely died out. I say "luckily" because the coca tended to dull a person's senses to the point where he became accident prone.

The true Indians had the high cheek bones, Mongol eyes and large banana noses. However, sometimes there was a lot of Spanish blood in them, so although I never saw a blond Indian, they often looked very European. They took a great delight in giving all the gringos nicknames, and their selection was very appropriate. A man with a pronounced limp as known as "Flat Tire,"another who carried his head tilted slightly to one side, due to a childhood injury, was known as "Five-to-Six." Other names such as "Prune Face,""Sucking Pig,"and "Cacique" (Big Chief) were more obvious. One old miner was known as "Cantinflas" after the famous Mexican comedian with the drooping trousers. Not seeing any resemblance, but realizing there must be a reason, I asked for the story behind the name, and was told the following.

Once a year the police would come around to the mines and villages looking for young men who had not done their military service, any youngsters they found were shipped off to the nearest army barracks and conscripted. The young miners objected strongly to this practice and often would hide out in the mine. The older miners would bring them food and tell them when the coast was clear and they could come out. It was during one of these visits that the police, without the knowledge of management, decided to fill their quota by making a raid underground. At the time the police arrived at Cantinflas' working place he happened to be relieving himself, but when he heard his helper yelling for help as he was dragged towards the surface by two policemen, Cantinflas grabbed an iron bar and went to the rescue, unfortunately he had forgotten to fasten his pants securely. Although he effected the rescue, his pursuit of the frightened police was brought to an abrupt halt when his pants fell down, tripped him up and sent him sprawling into a water-filled ditch. From then on his name was "Cantinflas."

Old "Popeye" was our senior shift boss, a village elder who had spent several years in the Peruvian submarine service, and had been discharged as a petty officer. During the course of his service he had visited both the States and Europe, but had finally returned to his remote mountain village and gone to work in the mine.

Most of the shift bosses and some of their men were making great sacrifices to put their sons through college as engineers or geologists. Some of these students received Company scholarships, and later completed their studies in the States. They generally turned out to be knowledgeable and useful men to have on the staff as they knew the Sierra and had good mining backgrounds.

Chapter Four
Andean Diversions

One of the more common diversions among the Andean Indians is a "pachamanca" which is a picnic with drinks, but with a difference. Long before the party is due to start, the organizers dig a hole in the ground and light a large fire close by. In the fire they heat rocks which are put into the hole when they are really hot, along with food wrapped in leaves, and on top of this wet straw or hay is piled, and then everything is covered with earth.

The mound is then left alone until the maestro in charge declares the food to be cooked, at which time it is dug up and shared out. The modern housewife with her oven thermometer, meat thermometer, kitchen scales and detailed instructions from a cookbook may well be puzzled by the maestro's ability to know when the food should be dug up. I guess the pachamanca chefs just learn from experience.

On arriving at a pachamanca, the guest is offered beer and sometimes wine (a very sweet variety) and always "pisco." Pisco is a colorless brandy made in Peru, and can be very good indeed. The Indian miners' pisco was rough and strong, and probably not aged more than a few days, if not just hours. However, it certainly helped to keep the cold mountain air at bay, and get everyone into the party spirit in a hurry.

Aguardiente or "caña" is something else again, made out of sugar cane, almost pure alcohol, and should be avoided. The miners took a great delight in offering a new young supervisor a real man-sized snort of the local white lightening. They would stand back and grin as the victim's eyes bulged, and he choked and gasped a "Muchas gracías!"

The meat at a pachamanca consisted of lamb or young goat, and was eaten with a hot chili sauce—something else the newcomer learned to respect very quickly. Potatoes and beans in the shell also accompanied the meal along with a sweetish corn paste cooked in the corn husks, sometimes containing some raisins. Peru is the home of the potato and there are many different varieties available. The yellow kind and a purple variety were served most often.

A delicacy cooked at a pachamanca is "cuye" or guinea pig, which is very similar to dark meat chicken in taste. Many a young company wife at her first pachamanca has been horrified to suddenly discover that the chicken leg she has been enjoying ends in claws similar to that of a kitten. This, plus a precipitate introduction to pisco and hot chile sauce, has prejudiced many gringos against pachamancas for life.

Sometimes along with a pachamanca there will be a "cortamonte," literally a tree cutting. A tree, fifteen to twenty feet high, usually a eucalyptus, is obtained and the butt buried in the ground so the tree stands vertical in the center of a flat area suitable for dancing. The tree is then decorated with colored paper streamers, and has some presents, bottles of aguardiente or pisco tied up amongst the branches.

Dancers gather around the tree, the women wearing many petticoats under their calf-length skirts giving them a bulky appearance. Over their blouses they would wear colorful embroidered waistcoats of satin with sparkling sequins sewn in patterns, on their heads broad brimmed high crowned white straw hats with black bands. The men would wear their best suits, always in dark colors, and broad-brimmed black hats.

The orchestra, violin and harp with two or three wind instruments, strikes up a huayno, and the dancing starts. The couples shuffle and stomp their way around the tree in a circle, and by the tree in the center of the dancing circle, the padrino responsible for throwing the party, stands with his partner. The padrino

Brian and friends starting out for a cortamonte.

carries an axe and his partner a bottle of pisco. The couple dance and gyrate inside the circle of moving dancers looking for another couple to start the ball rolling. When they have selected a pair, they offer them each a slug from the bottle and then give them the axe. After a couple of turns around the tree, the man with the axe takes one chop at the tree and then hands the axe back to the padrino, rejoins his partner and the moving circle of dancers. Turn and turn about, the dancing couples take their drink and chop at the tree until it falls over, and this can take hours.

At my first cortemonte I had the mistaken idea that the objective of the whole thing was to cut the tree down. It was only afterwards that I discovered that the one who makes the final blow is the guy who foots the bill for the next cortemonte! I also found out shortly after this, that hiring an orchestra, buying and setting up a tree and supplying all the free booze could run well over a hundred dollars, which was a lot of money in those days.

However, I was more fortunate that a large German friend

Participants at a cortamonte.

who was talked into giving his mine section a pachamanca and a cortemonte after they had broken a production record. As he was paying the shot, he was given the honor of first chop at the tree. Being a powerful man and not wishing the dancing to go on for too long, he gave the tree a pretty good clout. The shock dislodged a bottle of caña from the upper branches which fell and broke on his forehead. He was removed unconscious to be stitched up, and still bears to the scars to this day.

Cockfighting is not so common in the Sierra as it is on the coast, presumably it takes a fairly well-off community to have cocks to spare for sport killing. As each cock is armed with one long needle-pointed, razor sharp steel spur, the fights tend to be short and bloody, always with one dead bird and often with the winner dying shortly after the fight.

A spectacle which is still practiced in remote areas is the contest of the condor and the bull. After capturing a condor, the villagers sew its feet onto the back of a bull, and then let the two loose in an arena and watch the "sport."The bull, crazy with pain, tries to buck the bird off. The condor, balancing on its wings, starts pecking way at the bull's back in the region of the kidneys. If the bull has the foresight to roll, he kills the condor, if not, the condor will eventually kill him. Usually the condor is released before this can happen, and given a reward of a drink of chicha (a beer made from maize) for its show of valor.

I had two friends who were present at one of these affairs in a village some distance from camp. Shortly after it started, the condor managed to break free and fly away. When this happened, the crowd, which contained the average proportion of fiesta time drunks, turned ugly and decided that it was the presence of the gringos that had ruined the day. My friends departed rapidly, followed by a shower of rocks, and were lucky to escape without serious injury.

In the south of Peru, they have two sports that are native to those parts. Bullfighting there means pitting two bulls against each other, and large amounts of money are wagered on these contests. The bulls are quite docile with their owners, who often ride them into the field of battle. If the bulls appear reluctant to fight, a cow is led in between them. Having a lady to fight over is always enough to start some action — bulls are not that different from humans! The bulls do not kill each other, seldom really hurt each other, one eventually gets tired and leaves the field to the victor.

The other southern sport is pretty rough. The menfolk of two villages will agree to meet at a certain time and place, armed with rocks and slings to fight it out. These slings are basically the same as the weapon David used to kill Goliath, and in the hand of experts can be lethal. No protective devices such as hard hats are allowed as it is considered sissy. Any number can play and there is no time limit, the fight continues until one side quits. The winners of this mayhem then help themselves to all the single girls of the defeated village.

One old custom is played with a duck. The duck, often dressed in doll's clothes, is suspended by a rope slung over a branch of a tree. Riders then gallop toward the duck and try to grab it, while an Indian on the other end of the rope tries to jerk the duck out of the rider's reach. If the guy on the end of the rope gets careless or tired, it's pretty rough on the duck who loses some feathers before finally being captured.

On one occasion the duck had the first laugh, if not the last. The duck was swinging gently on the end of the rope quacking nervously, and wondering what it was all about, when the first rider of the afternoon came pounding down toward him. The horseman stood in his stirrups straining upwards as the duck was jerked from his reach, but the sudden jerk caused the frightened duck to empty its bowels, which caught the passing rider full in the face. This was the kind of action that always appealed to a local crowd, and it had them rolling in the aisles.

Nowadays it seems that football, volleyball, basketball and "futbalito" are the games people are most interested in. "Futbalito" is football (soccer) played on a small scale, the field being about the size of a tennis court with smaller goals, and only five men per side. Full-sized football, though, is the most popular of all the games, and is played with much enthusiasm, notwithstanding the plus 14,000 feet altitude of some of the soccer fields.

One of the features of these mountain villages are the buses which ply back and forth on the narrow mountain roads carrying goods and people. Their proud owners paint them with bright colorful designs and print a name or slogan on them as well. I once saw "God is My Guide" sitting in the Mantaro River after missing a curve, I should hope the owner changed the name after this misadventure.

The owners pack in as many seats as possible, so that after a short trip, if one is of normal height, both knees will be skinned by the seat in front. One can pay more and have more leg room sitting in front alongside the driver. However, in this position the traveler needs nerves of steel, as he has an uninterrupted view of the road and can observe the driver at close hand trying to overtake on corners, or switching the engine off and coasting when the road is downhill.

The object seems to be to drive as fast as possible and overtake everything in sight. Most roads are unpaved, rough and very dusty, and this form of travel, complete with chickens, a pig or two, sacks of farm produce in the aisles and a full complement of cheerful Indians can be quite an experience.

Once traveling on a packed bus, our driver was trying desperately to overtake a bus from a rival company. Almost blinded by dust, we were tailgating the bus in front when we arrived at the entrance of a narrow tunnel. As the driver started to brake, the bus caromed from one side of the tunnel to the other, shedding broken window glass over badly scared passengers. When we finally left the tunnel behind and stopped, the passengers rose as one man and grabbed the driver and threw him out on the roadside where they proceeded to beat him up.

Before he got too badly hurt, I tried to tell my fellow passengers that if the chauffeur was disabled, there would be no one to drive the bus, and we would be stranded. They listened to reason when I pointed out that the logical thing to do was to beat him up after we arrived at our destination.

Muttering dire threats, they pushed the battered man back into his driver's seat and everyone climbed aboard. We took off slowly, and for the rest of the trip the chastened chauffeur had thirty-six voluble backseat drivers to contend with.

On a night trip these buses could be very cold and drafty, as no heating was provided. One thoughtful driver, with comfort of his passengers in mind, connected a hose to his exhaust pipe and poked the other end up through the floorboards into the main body of the bus. It was only after passengers started to fall off their seats that the driver stopped to see what was going on. Unfortunately, it was too late for one of them.

Company hospital at La Oroya.

Chapter Five
Safe Conduct

The Safety Department was the smallest department in any of the Cerro mines, but it reported direct to management and handled one of the most important functions in the whole operation. When you have several hundred men working underground in separate working places where supervision cannot be constantly exercised, the men themselves have to work safely.

When a man first started to work underground, he was usually nervous in his new environment, but listened carefully to the supervisors and the safety engineer's instructions. This group might have silly little accidents such as hitting their thumbs with a hammer or dropping a rock on their feet. As the new men would start out as helpers with experienced miners as partners they had someone working alongside to keep them out of serious trouble.

When a man had five or six years' experience and had just been promoted to be "maestro" or lead miner in a working place, he was most likely to have a serious accident. By this time he figured he knew all the answers and would take risks. Most of the serious and fatal accidents in our experience occurred to this group of workers.

The three main causes of accidents underground were things falling on people, people falling, and accidents with explosives. Things falling on people could be anything from a single rock to several hundred tons of rock coming down in a major cave-in. In general, if a person is buried by fine material there is very small chance of saving his life, as he suffocates. We have been able to dig a man out in three to four minutes in some cases, and still lose him. If a man is buried by large material he might get banged up pretty badly, but if he can breathe, there is a good chance of getting him out alive, even after several hours of entombment.

A friend of mine was buried for several hours in a mine in Canada and lived to tell the tale. It was coarse material and he could hear his rescuers digging him out and even managed to talk to them. However, he had been buried with arm outflung, and the pressure of the rock had cut off his circulation so that he had no sense of feeling left in any of his limbs. As the Canadian miners were digging away they would make remarks and pass on advice of their progress. Finally one hand was uncovered, and one of the rescuers grabbed hold of it and said, "O.K., Mac, it won't be long now, I've got hold of your hand, feel it?"

As mentioned, the victim's circulation had been cut off and he couldn't feel anything. On informing his rescuers that he could not feel anything, he heard one of them remark to the others, "Jesus Christ, his arm's off!" This did nothing to help his morale until such time as he could see that all his limbs were still attached. He got out of this badly bruised and with a lot of stitches, and for a while afterwards, some pretty vivid nightmares. However, he also had a good story to tell at a party, and of this he took full advantage.

One other story about this engineer deserves telling, he is a person to whom things happen, and so has many stories associated with this career. As he is very sharp-witted, with a good command of the English language, it is very seldom that he does not have the last word in any verbal exchange. One day while driving up to the mine in a clapped-out old pickup truck, I was giving him a line on the work to be done that day. He was listening intently when I swung the vehicle around a hairpin bend. As we went into the curve his door opened and he went out the cab hanging on the door. As I straightened out, the door swung shut and deposited him back on the seat. Glancing over and noticing his white face, I pointed out to him that it was extremely rude to leave like that before I had finished speaking. For once I had the last word!

The problem with trying to rescue someone who has been caught in a major cave-in is trying to calculate if there is going to be any further movement which could endanger the rescue crew. This risk has to be balanced against the chances of finding the victim or victims alive. The Indian miners were impetuous, and completely disregarded their own danger when embarking on a rescue operation. They would even take appalling risks to rescue equipment from a caving working place unless restrained.

We had two men crossing a heavily timbered area on a level when two working places caved below them. The tunnels in that area dropped and left a mass of twisted timber and broken rock. The chances for their survival were practically nil, and the floors of the tunnels leading into the area were cracking, indicating the possibility of another drop. To try to support a dangerous floor in a case like this is impossible (a bad roof is not too much of a problem). In spite of this, the crew of experienced miners we assembled to carry out the rescue, who well knew the danger, could barely be held back until we could check to see if the floor was still moving.

Trying to find a body in a major cave-in such as this can be a problem in itself. In a hot location, through, it is not long before your nose will be able to guide you.

People always seem to miss their footing and be able to fall, even on a level surface. Although, as one would expect, the seriousness of the injury usually depends on the height a person has fallen, this is not always so. We had a man fall ten feet and break a rib which pierced his kidneys and killed him, and another fall fifty or sixty feet and get away with only scratches. The worst accidents are generally those associated with explosives. If the person isn't killed, he is usually fearfully maimed.

I will always remember the great courage exhibited by one victim of a blasting accident. The maestro and his helper were driving a tunnel and had drilled some fifty holes. These they had loaded with dynamite and as it was the end of the shift, proceeded to light the fuses. It was against our regulations for one man to light this many holes. The smoke from the burning fuses makes it very difficult to see and a man could delay too long lighting the fuses so that he might get caught in his own blast. Furthermore, we used a "spitter" for lighting the fuses, rather like a firework called a "sparkler" by children, and when one spitter had been exhausted, a miner was instructed to leave the working place regardless of whether he had managed to light all the fuses or not.

For some reason, the maestro neglected both these precautions and the first holes exploded while he was still attempting to light the remaining fuse and killed him instantly. His helper was further back and received part of the blast in his face and lost both eyes. He picked himself up and started heading down the tunnel to get out of the danger zone when he was hit in the back and knocked down by the next holes exploding. In spite of this, he managed to make his way down the tunnel to a manway which went to the level above, where the men collected at the end of the shift before leaving the mine. Leaving a trail of blood, the helper climbed twenty-seven floors of ladders, about one hundred and eighty feet, to the upper level to get help for his maestro. All this with no eyes and not much of a face left.

When holes are blasted to break the rock, sometimes they do not break right to the end and this remaining socket is known as a "bootleg." Sometimes these bootlegs will have some remnants of dynamite in them. Consequently, after blasting it is important to wash down and inspect the new face before you start drilling again to make sure you don't drill into a bootleg containing powder and have an accident. A miner who was killed by doing this in Canada was reported in the local newspaper as having hit a bootlegger who promptly killed him!

Nitroglycerine base dynamites can give you very bad headaches when you handle the sticks to load your drill holes, because the nitroglycerine goes through the skin. If you have tough horny hands and rinse them off after loading your holes, you will probably be alright, but for a newcomer who has not yet acquired enough calluses it can be very unpleasant. It is not unknown for a person who is disliked to have some small pieces of dynamite put in his boots. Until the nitroglycerine is all used up or the boots thoroughly washed, he will have headaches that feel like they would crack the skull of a rhinoceros.

When blasting, the crew responsible should always guard all approaches to the danger area. It is useful to have a good nose, and if you smell smoke from burning fuses, proceed with the utmost caution. In one case we had a miner relieving himself in an abandoned work area at the end of shift. Intent upon the job at hand, he didn't realize that a long diamond drill hole connected his temporary bathroom with a face that had been prepared for blasting. When the blast went off, the diamond drill hole acted rather like a rifle barrel pointed straight at the unfortunate victim's rear end. The shock blew him head over heels out of the tunnel and relieved him of any problems he may have had with constipation, at the same time leaving him with enough scars to remember to use the proper location in the future.

Sometimes a miner will take the opportunity of working underground to even the score with an old enemy; murder is difficult to detect under the circumstances. Several cases of accidents which looked extremely fishy were on our books, but nothing could be proven, and so the classification "work accident" was entered in the records.

Another responsibility of the Safety Department is to check the underground ventilation, both for temperature and humidity, and also for noxious gases and dust. In very hot humid areas people can easily collapse from heat exhaustion and should take salt tablets with their water. Gases have been discussed earlier, but dust is another story.

Rock which contains high quantities of silica produces a fine dust with sharp cutting edges that damages the membrane of the lungs. The resulting scar tissue on the surface of the lungs reduces the effective working area of the lungs and causes a disease called "silicosis." The victims also become very liable to catch other lung diseases. Silicosis was a terrible scourge of miners in the old days, particularly in the old Cornish tin mines and Witwatersrand gold mines of South Africa. Modern methods of control and routine checks of the mine air have done a great deal toward eliminating it. Yearly x-rays of all underground personnel enable doctors to detect the beginnings of silicosis early, and to remove victims from the mines so that they can work in uncontaminated areas on the surface.

Apart from the normal run of accidents, there is always the danger of a catastrophe claiming the lives of several score workers. Coal mine explosions are, unfortunately, well known in this category. In metal mines it is usually a fire or inundation of the workings by water or mud. The Sunshine fire in Idaho and the inundation of the Mufulvia mine in Zambia with mill-tailings are well known examples of what can happen. Our mines also had three bad accidents during their history. One was a shaft fire where over fifty men were trapped underground and gassed. Another was the inundation of a different mine when a working place broke into the bottom of a lake with the loss of over eighty lives, and the last was a coal mine methane explosion which claimed over fifty lives.

In the event that a fire or flood appears to be getting out of hand, it can become necessary to swiftly evacuate everyone from the mine. At first this appears to be a terribly difficult problem as it is not unusual to have several hundred men working in separate pairs spread over miles of workings and there may be 1,000 feet or more between the lowest level and the highest. The system used is old, but the most effective that anybody has yet been able to come up with and also simple to operate.

Every working mine uses compressed air for its rock drills and often for hoists and other equipment. In this safety check a foul smelling gas called ethyl mercaptan is introduced into the main air supply on the surface, and within minutes is making its presence felt in all the working places. We would run these tests about once a year to make the men familiar with the system, and also to see how fast we could evacuate the mine. It was not unusual during one of these exercises to find some inexperienced Indian miner carrying his drilling machine with him to the workshop. Having only half understood what was going on, he had thought there was something wrong with his equipment and that it needed attention.

In one case the ethyl mercaptan nearly caused an accident in itself. A routine evacuation check was to be carried out and the mine superintendent was standing by the compressor house waiting to give the order to inject the ethyl mercaptan at the correct time. Unfortunately, the after coolers on the compressor were not working correctly, and the compressed air was too hot. When the mine superintendent gave the word, there was a terrific explosion and several hundred feet of the main compressed air line disappeared into fragments. Luckily, no one was hurt although the mine super was narrowly missed by a 15-foot section of 24-inch pipe flying past his head. After this, all mines modified their warning systems to prevent any possibility of a recurrence.

Besides the Safety Department, a mine would have the normal service departments such as Accounting, Warehousing, Industrial Relations, Maintenance, etc. The Head of the Maintenance Department had to be a very talented individual as he had to have a broad knowledge of all the equipment in the camp, and be able to fix the mine superintendent's washing machine or freezer, and at least once was called upon to repair the cathouse jukebox which belonged to the local priest. The priest was most insistent that this problem should be given priority, as every day the jukebox was out of action it was losing him revenue.

Clown at annual rodeo at the Casaracra Station.

Chapter Six

Minerals and Metals Deposits

The Concentrating Department receives the ore from the mine and upgrades it and sends this "concentrate" to a smelter. For instance, a mine might produce 5,000 tons of ore per day grading 2% copper, which means it contains 100 tons of copper metal. Say the concentrator recovers 90% of this, or 90 tons in a concentrate running 25%, then the final product sent to the smelter would be 30 tons of concentrate running 25% copper. Obviously, the savings in transport costs and smelting costs easily offsets the loss of 10% of the metal in the treatment stage. There comes a point, however, when dealing with very rich ores, whereby the pendulum swings the other way and it is smarter to skip the concentrating stage and ship direct to the smelter.

In all treatment processes the first step is to reduce the size of the ore. Some mines, where labor is cheap, might hand pick the waste rock out of the ore from a slow moving belt before delivering it to the concentrator. However, crushing is an essential first step to the concentrating process. The ore is generally reduced down to about 3/8 of an inch by passing it through a series of crushers and screens set at different sized openings. To reduce it still further in size, it then passes into the mills, and water is added.

A mill is like a large horizontal cylinder of steel containing steel balls or rods, and which revolves at about forty revolutions per minute, the speed will depend on the diameter of the mill and various other factors. If you went too fast, centrifugal force would hold everything on the side of the mill. The ore is fed into one end of the mill, and the cascading balls, or rocks, reduce it to a mixture of sand and slimes.

The object of this reduction in size is to physically break the ore particles free from the host rock in which it is embedded. As all ores have different characteristics, the amount of grinding necessary will vary. In some cases, such as the tungsten ore called wolframite, the ore is very friable, and too much grinding will reduce the wolframite to such a fine state that a great deal of it will be lost and recovery will be low.

In some mines the grinding is done dry, but this is not a common practice as water is generally used in the grinding to product a slurry. Very large quantities of water are necessary, a tin ore may need as much as 7 tons of water per ton of ore treated. For flotation treatment of sulfide ores, 2½ to 3 tons of water per ton of ore is a more normal figure. In desert country, where water is scarce, this figure might be reduced by recycling to about one ton of water to a ton of ore treated. This is still a lot of water when you consider that it is not unusual to find mills treating 30,000 to 90,000 tons of ore per day, seven days per week. You are then thinking in terms of 20 million gallons of water a day.

After the grinding stage, the treatment of ore will vary depending on its constituents. The most common treatment, especially for sulfide ores, is flotation. It is said that this process was discovered when someone watching a miner's wife washing her husband's clothes, noticed that the small particles of metal-carrying mineral attached themselves to the soap bubbles and floated free of the waste rock which sank. Whatever sparked the original concept, the process essentially is just that.

By adding chemicals and agitating the slurry from the mills and introducing air, a foam is produced which carries off the metal-carrying part of the ore. By using different combinations of chemicals, and by using a series of cells, the mineral can be retreated to increase the value by concentration and cleaning, or different minerals can be separated, such as galena (lead sulfide) from sphalerite (zinc sulfide). The concentrate will then pass to thickeners and filters to remove the excess water before it is sent to a smelter.

The "tailings," or waste product, will also go through a thickening process in order to recover as much water as possible for re–use, and is then sent out by pipeline or flume to a disposal area called a tailings pond or tailings dam. These tailings dams are the biggest man-made structures on earth, containing hundreds of millions of tons of material. They are called structures because you cannot just deposit these large quantities of waterlogged sand and slimes in a haphazard fashion, the system of deposition has to be designed and planned.

For one thing, uncontrolled deposition would take up very large areas of valuable land. If you pile it up or deposit it in a depression behind a dam, you have to be sure that it will stay put, and that storms or earthquakes will not produce a sudden avalanche of mud that will destroy all in its path.

Not long ago a tailings dam let go in Chile and wiped out a village lower down the valley causing over a hundred fatalities. These tailings deposits have to be built up very carefully, and kept well drained so that they won't liquefy and collapse.

In arid areas the dust from these dams can cause problems, especially if they are near a town or city. Nowadays, a great deal of work is being done to plant vegetation on tailings dams both to prevent dust and to bind the tailings together by means of the plant roots. This program also satisfies the ecologists, in that a shrub-covered hill is certainly more aesthetic than a big pile of sand.

The oldest form of separating metal-bearing minerals from their waste rock is by taking advantage of the difference in specific gravity. One cubic inch of cassiterite, a tin ore, is very much heavier than its host rock. The old time prospector, panning for gold, used this process by swirling his pan and gradually washing away all the lighter particles and just leaving the gold behind. By using sophisticated equipment, a gravitational mill will do essentially the same thing.

For iron ore and some other ores that have magnetic properties, high intensity magnets are used to effect the separation of the valuable mineral from the waste. Gold ores are mainly treated by cyanide which dissolves the gold. The solution, after it has been clarified, is treated with zinc dust, and the gold precipitates out as the zinc replaces it in the solution.

Another system is hydrometallurgy, which is under intensive investigation at present. First used on a large scale in the treatment of uranium ores, it is basically the chemical separation of the mineral, and sometimes the metal itself, by leaching the ore with acids or other chemicals. In the case of copper, it is possible to produce the end product, i.e., copper metal, right at the concentrator. At the moment, you have the right ore to start with, preferably a copper oxide or sulphate, but pilot plants are already operating and testing the treatment of sulfide ores.

The possible attractions of hydrometallurgy are many. The goals of the researchers are to make it economically viable, and also to come up with a sufficiently pure product to be competitive. With copper, you don't have to ship your concentrate long distances to a smelter, and the process does not produce noxious smelter gases which have to be controlled in some way to comply with governmental pollution laws.

There are other specialized means of concentrating ores, and new ones are being investigated all the time. In the diamond mines they use grease tables, as diamonds have a tendency to stick in the grease and the waste doesn't. There is also a colorimetric system which can separate different ores and waste according to their colors.

Concentrate from the concentrator is shipped to a smelter which will produce an impure metal and then, by means of its refinery, produce the final pure product ready for use by industry. Smelters use pyrometallurgy, which means treatment by fire. The ore, according to its constituents, is blended with fluxes and then roasted. In the case of sulfide ores, this is where the pollution problem comes in. The roasting of the sulfides produces large quantities of sulfur dioxide gas which can kill off vegetation over large areas surrounding a smelter.

Systems to control this gas emission have been in use for some time, but they are an added expense to operate, expensive to install in the first place, and not all of them can meet the latest pollution control specifications. On top of this, if the byproduct of the gas control system is sulfuric acid, which is by far the most common by-product, what does one do with the thousands of tons of sulfuric acid produced? High strength acid in itself can cause a disposal problem if it cannot be absorbed by local industry. It is this pollution problem which has sparked intensive research into the possibilities of hydrometallurgy.

There is no doubt that these added expenses, crash programs for pollution control and crash programs for research, not to mention delays in smelter construction and expansions awaiting legal and research answers, are in turn curtailing production and causing shortages contributing to the scarcity and high prices of metals which in turn spur inflation.

Metals are sold according to the market price and according to certain specifications of purity. It is impossible, except in the case of alloys, to come up with a better product than your competitors and to have a worldwide patent protecting your product such as manufacturing companies can obtain. The price you get for your product depends on supply and demand, and your competitor is in the same boat as you are. The difference between your profit and his will depend on the richness of your mines and the efficiency with which you control your production costs.

If pollution laws in your country or state are more stringent than those ruling your competitor, you are at an immediate cost disadvantage. In times of surplus, your competitor can always undercut your prices and make a profit while you register a loss.

One suggestion put forward is to build your smelters where pollution controls are not yet in effect. This is not, of course, a very feasible solution. In any case, pollution control is becoming popular worldwide. Recently, the government of a developing country in Africa was asked what standards of pollution control should be used for a projected mine-smelter complex, and was given examples of existing regulations from other countries to choose from, as there were no existing regulations.

The location of the project was in a desert surrounded by miles and miles of nothing. Much to the design engineer's consternation, the government chose the most stringent of the regulations that had been provided for study. No doubt keeping up with the times is an admirable desire, but the impact of control legislation brought out too hastily, and without proper long range planning, inevitably and lastingly makes itself felt in the pocketbook of the general public.

To conclude with the smelting process, after the sinter is produced from the roasters, it then passes to another set of furnaces. This stage varies considerably, depending on the metal being treated, and the particular smelter and its choices of equipment and process system. It is after this step that the metal first appears in recognizable form. In the case of copper, it is called "blister copper" because the surface of the metal looks blistered.

To purify or refine the metal it goes to a refinery which generally uses a process of electro-deposition. The metal is cased into anodes which are in the shape of flat plates with ears at the top so that they can be suspended in special tanks. In the tanks there is an electrolyte, a solution which will carry an electric current, in copper refining, copper sulfate solution is used. In between the anodes, but not touching them, starting sheets are placed. These consist of very thin flat plates with hanging lugs at the top, and are made of pure metal.

The starting sheets are made by electrically depositing the copper onto stainless steel plates and then peeling them off when they have reached the required thickness. When an electric current is passed through the anodes, electrolyte and cathodes, the copper goes into solution in the electrolyte, impurities in the anode dropping to the bottom of the tank as a sludge. The copper follows the current flow and deposits on the starting sheet, gradually building it up to form the finished product or cathode. The sludge containing the impurities will often contain gold and silver which can be recovered by further separate treatment.

Often the cathodes will be melted down to form wire bars before they are shipped, as this is a more acceptable product for industry, and commands a higher price. It is also a handier form for handling and shipping.

Iron ore goes through no refining process as the final product is produced from a blast furnace. The concentrator will upgrade lump iron ore by magnetic separation. In order to obtain a higher grade product, the pelletization process is becoming more and more popular. This also has the advantage of enabling the mine to improve the product from low grade ore bodies. The ore is crushed and reduced to a slurry by using mills, and the iron ore separated from the waste by magnets. This fine product is not suitable for blast furnace use and so is pelletized by mixing with bentonite, a binding agent, and then roasted to produce a hard pellet about ½ inch in diameter.

As iron ore is shipped in large tonnages, very often over great distances, the higher the grade of your product, the less money you are paying for transporting waste. You are only paid for the iron content of the ore. Pellets can grade in the high 60's to low 70's in percentage of iron, and now semi-reduction is being tested to try and improve on this.

Clubhouse of the Rod and Gun Club, Lake Pomococha.

Chapter Seven

Camp Pastimes

Living in these isolated mining camps promoted some pretty heavy drinking, but everyone always went to work on time the next morning and sweated out the alcohol, and this seemed to prevent any real bad aftereffects. I can only think of two or three isolated cases where a man broke down and just crawled into the bottle and pulled the cork in after him, so to speak.

In one camp where we were partying even more often than usual, an epidemic of hepatitis struck a number of the non-drinkers. This put extra work on the drinking fraternity, who nevertheless still made it to work on time. I often thought afterwards that we probably had hepatitis too, but felt so lousy every morning anyway that we didn't realize we were sick.

Fishing and hunting were two great pastimes. Rainbow trout had been introduced to the watersheds back in the early part of the century by members of the mine staff, and trout were, and still are, abundant around most of the mines. The days of twenty and thirty pounders were gone by the time I got there to fish, but eight to twelve pounders were not uncommon – if you knew where to go. The Indians became more sophisticated in their fishing methods with better fishing gear, nylon lines and treble hooks, etc, but it was indiscriminate dynamiting of the trout which has done the most damage.

Hunting was mainly for partridge on the altiplano grasslands, or for the various types of duck available on the lakes. Both kinds of hunting were fairly strenuous on foot at altitudes of 12,000 to 14,000 feet. Even more exertion was involved in hunting a bird called a francolina, which is similar to a partridge, but two to three times the size, which lives in the real high country. They would go around in flocks of about a dozen or so, often in areas of 15,000 feet or more, and usually it seemed in very rough terrain. By the time you got within range you were puffing so hard from the exertion that it was impossible to aim straight.

Apart from the exercise and diversion, the game and fish made a welcome change to the menu. In the early days, the food in camp, particularly for the single men, was pretty poor to start with, and the cooks made it worse. For a long time our camp boasted of a cook who could get grease into a boiled egg.

Most of the mine superintendents of that time had come up during the depression of the '30's, and they did not believe in pampering the staff or the men. Generally, a complaint or suggestion for an improvement was met by the remark that a train came up the hill every day and one went down, and if you didn't like it in camp, leave. They had been brought up through an exceptionally hard school, and had been taught to watch every penny. When it came to spending money on anything that didn't produce ore, my first super, for example, was known as being "tighter than a bull's arse in fly time."

The loneliness and boredom in some of the single men's hotels or bunk houses would occasionally affect people in funny ways. Usually, they would suddenly get an urge to start breaking up the furniture or smashing windows. Providing the damage was paid for, this was not considered as unduly getting out of line. In one large bunkhouse which had plasterboard walls between the rooms, they would stage a race from one end to the other through the partitions. Not everyone could take part as it took a lot of "avoidrupois" to get up enough momentum to smash through a partition with only a 10-foot start.

One champion was doing fine, and was halfway home when his forehead came in contact with a 2x4 strut hidden behind the plasterboard. He stepped back, lowered his head, and to

Brian with Dick McCombe and son setting out for a hunting trip.

the credit of his Marine Corps training, made it to the end of the course in spite of the blood pouring from his head wound. Only two races could be run at a time, one each side of the central corridor, as considerable repairs had to be completed before the partitions were back in shape to provide suitable obstacles.

Bicycle-polo down the bunkhouse corridor was a short-lived pastime in one camp. It was short-lived because the bicycles couldn't take it, and the players finally ran out of steeds.

The bigger camps had bowling alleys and golf courses, and would arrange intercamp golf tournaments which usually ended up in some wild parties. The visiting team would be feted and partied into the early hours of the morning while the home team was sent to bed early. During the match, drinks would be served at various well sited locations, and your caddy would hold your gin and tonic or rum and coke while you drove off. It took pretty robust players to make eighteen holes, and it was not unusual to see a visiting player being carried off the course.

The courses were nine holes, with differently placed tees for the second nine. On some of the courses the rough was a mean type of tufted grass that grew two to three feet high. During a strike we once decided to use the mine equipment which was idle, to deal with the rough once and for all. Unfortunately, the course was in a hollow and the ground was none too firm. Before our volunteer operators reacted to the danger, we had several large pieces of earth moving equipment mired down in various parts of the course.

The following day being a Sunday, we did not have time to remove all the stuck units before the usual Sunday morning golf games would start. A hasty meeting of the rules committee was called, and it was decided that if the D9G tractor in the middle of the No. 5 fairway impeded your shot, you could move the ball without penalty, but should not place it closer to the hole. The two D8 tractors out in the rough, though, were considered natural hazards, as it was pointed out you shouldn't be in the rough anyway. The various craters in the fairway from which stuck equipment had been removed were considered "ground under repair." That day was the only time I have ever registered a negative distance with a drive, when my ball, slightly off course, struck the blade of a D8 and rebounded behind the tee, narrowly missing my opponent in the process.

At another camp, we were plagued by a type of large goose called a "huachua" defecating all over the greens. It so happened that one Santa Rosa's Day, she was the patron saint of the police force, it was a Saturday, and we had all four members of the local guardia in the club celebrating their saint's day. Then, for some reason, the goose problem came up in conversation. By this time, none of the party was feeling any pain, and one of the policemen asked me why we didn't shoot the offending birds, and that they would certainly help us in the project.

No sooner said than done, and we all dispersed to gather arms. The camp at that time had a small armory of old lever-actioned Winchesters which fired a .44 lead slug. They had short barrels, kicked like mules, and after a few rounds you couldn't see anything for the smoke. We also had a few privately owned shotguns.

The police soon returned carrying some WWI Manlichers and one German 9mm submachine gun, which belonged to the sergeant. Much to my joy, I managed to talk the sergeant into swapping his submachine gun for my 16-bore shotgun. Suitably armed, already well primed, we headed down the road to the golf club.

Sure enough, the offending geese could be seen on several of the greens showing their usual contempt for Sunday's golfers. We split into parties to attack each group of birds simultaneously, and like a group of raiding commandoes, deployed across the rough and fairway. Silently, apart from the occasional hiccup or beery belch, we crept within firing range. Fire was opened together with a loud roar, clouds of smoke and long muzzle flashes coming from the old Winchesters, with chunks of sod flying out of the greens from the bullets, making a scene reminiscent of a Civil War movie. Finally, quiet returned to the battlefield, and we discovered that in spite of the fire power brought to bear, not one goose had been hit, and luckily, no people either. However, three greens and two flags had been shot to ribbons, and also one bench on a nearby tee had taken a burst from the submachine gun. True, we didn't kill any of the birds, but it was at least two weeks before they came back to annoy us again.

Visitors were always welcome in a camp. A new face and someone different to talk to made a nice change. After a certain amount of time all living together, it was inevitable that eventually people would start telling your stories back to you as their own. This was irksome to say the least, consequently, a visitor would be entertained royally. One visitor who probably remembered his visit for quite a while was the Bishop for the west coast of South America.

His first afternoon in camp he was invited to tea to meet all the wives at one of the houses. The tea party started on time, and was proceeding quite nicely, the ladies and the Bishop clasping delicate cups of tea with small plates and napkins balanced on their knees, when the day shift ended.

The hostess's husband plus the husbands of two of the ladies present arrived from work, and after meeting the Bishop and chatting briefly, the men disappeared into the kitchen. During the next half hour most of the other husbands arrived and after greeting the Bishop also disappeared through the swinging doors into the kitchen. After a short while it was noticeable that the conversation in the kitchen had increased in volume, and that some of the language was getting out of line. It was not too much later before a fight started and one of the combatants came flying backwards through the swinging doors to land on his back at the Bishop's feet. With a few choice expressions, the recumbent figure scrambled to his feet and rushed back to the fray.

At this point the ladies escorted the Bishop from the scene of action back to the house where he was to spend the night. Meanwhile, peace returned to the party, new men arrived and it went on into the wee hours with the singing of bawdy songs. It was not until the next day that people realized the Bishop's bedroom had been next door right across from the party. He never mentioned the serenade, but he never came back for another visit either.

Cards were a popular pastime. Families played bridge together, but when the men got together they preferred cribbage or poker. Poker was often dealer's choice, and variations such as "spit-in-the-ocean," "rock crusher," "night baseball," "seven-toed Pete," and "low-in-the-hold wild" would leave a newcomer not only confused, but usually somewhat the poorer. Liar's dice and "dudo," another Peruvian dice game, were favorite ways of deciding who should pay for a round of drinks.

In a Canadian construction camp, employing some 2,000 men, card games were not allowed in the bunkhouses because of the fights and damage that used to result from the altercations over the cards. Instead, special tents were erected and the card players had to use them for their games. After pay day, twenty to thirty separate games would finally end up in the hands of four or five players, and then the big game with several thousand dollars would take place.

During one of these evenings two very large tunnel men got into an argument and eventually a fight broke out. As they both weighed around 240 pounds, it is not surprising that not only the tent was wrecked, but also all the furniture. However, one of the other players or one of the kibitzers had taken off with the pot during the confusion, much to the combatants disgust.

At one Canadian camp where they allowed gambling in the bunkhouse, a game was progressing into the early hours and

"Balleyes" McFarlane and "Gimme-Gimme" Steve were finally cornering most of the camp pay checks. However, they didn't realize they had been interrupting the sleep of a large Finlander who finally decided to break up the party with a double-bladed west coast logger's axe. With the first swing he cut the electric light cord and plunged the room into darkness. As players took off through the windows and out the door, the Finlander set about the furniture and everything else within reach. It was reported that the next morning revealed the shambles, and some quarters scattered on the floor that were well and truly "two bits," as they had been neatly cut in half with the axe when the Finn had hacked the table in two.

In Peru, every Fourth of July would be celebrated with softball games, and sometimes one camp would challenge another. As these games took place infrequently, the day following a match would see most of the previous day's players limping around with pulled muscles and strained backs. Enthusiasm rather than skill was the order of the day. One famous game between two camps, played for a case of whiskey, ended up with the visitors winning 31 to 30, and four players hospitalized.

For variation, a camp would sometimes have a costume party, usually at Halloween. These always got off to a rousing start, as very often the men felt so stupid dressing up they would get well primed with booze before they even got to the party. One of the risks at this type of party would be an emergency phone call from the mine, perhaps a cave-in or a man injured. The superiors concerned would race up to the mine to get underground, often forgetting what they were wearing.

One surprised shift boss, tackling a ground failure in an underground work place, was shaken to the core when the mine superintendent arrived with two other foremen all wearing makeup. They had been at a party as the Andrews sisters, and had forgotten to remove their eye shadow and rouge.

Most camps had horses which the geologists used for exploration trips around the mines, and these could often be borrowed for fishing trips and riding expeditions during the dry season. The horses were small tough mountain ponies, and often not broken in too well. We would take off on a trip well clothed and draped in shot guns, fishing rods, cameras and packsacks with beer and sandwiches. When a rider fell off, which happened quite frequently, he would land with a loud crash. Those still mounted would then have to chase after the riderless horse and catch him for their grounded companion.

The saddles were very old U.S. Army issue. They had a split down the middle, which reputedly made them more comfortable for the horse. Anyone who has ridden one of these saddles for eight hours or more, after not riding at all for several months, will remember them vividly.

The dry season when we went on these excursions was also the middle of winter, and although the days would be sunny and cloudless, the nights in the high altitude could be bitterly cold. Camping out on a riding trip was rough, you had to keep your boots in your sleeping bag overnight or they would stiffen up and be impossible to put on in the morning. One can see why people who live in the altiplano, whose only source of heat is dried llama dung, and who have little or no hot water, don't bathe very often, if at all.

Old Lowdes Shaft, Cerro de Pasco.

Chapter Eight

Underground Accidents and Emergencies

To handle accidents and underground emergencies, the company had two mine rescue teams trained to use the various types of oxygen breathing apparatus. Each team, one of supervisors and one of shift bosses, consisted of six men. Five was the normal number of a crew, but we had one extra in case anyone was away on vacation or absent for other reasons. As two of the mines were subject to spontaneous heating of the ore, which could and did result in underground mine fires, these crews were essential, and frequently called upon to act in real emergencies.

Every month the crews had to prepare and put on their special equipment and go through an exercise such as building a ventilation stopping or rescuing an "injured" man and bringing him out of the mine on a stretcher, and for this we were paid a small bonus. The equipment we used most often was a McAu breathing apparatus. It had a cylinder containing oxygen under a pressure of 2,000 pounds per square inch, and a canister containing cardoxide, which could remove the carbon dioxide from your expelled breath so that the oxygen remaining could be recycled. The whole apparatus looked something like a frogman's self–contained diving apparatus, and weighed about forty pounds. Normal usage was for two hours, but by conserving energy, and thereby oxygen, it could be made to last up to eight hours — I never heard of anyone actually proving this though.

We also had a much simpler apparatus with just an oxygen cylinder and no recycling system, which lasted half an hour, and then another type which worked purely by regeneration. In this last type, you breathed through a canister which removed carbon dioxide and generated oxygen. As the canister heated and the supply of oxygen built up, two rubber bags would gradually fill up with a reservoir of oxygen to breathe. The trick was to get the thing working well before you went into a dangerous area, or it was likely to quit on you. Luckily, we never lost anyone using these chemical masks, although it was a close call several times.

The way to get an underground fire under control was to close off all the openings into the burning area, so that the fire could get no oxygen. In a big mine we would then pour strong concrete bulkheads around the area and isolate it completely, we would leave pipelines through the upper bulkheads and then pour tailings suspended in water from the concentrator through the pipelines into the fire zone. (Tailings are the fine sand and slimes rejected from the concentrator after the valuable mineral has been extracted.) As this tailings and water mix had a 1,000 feet or more of hydraulic head, it arrived at its destination under considerable pressure, and thus would fill all the openings and smother the fire. After some years, once the tailings had hardened, the area could be opened up and mined again. These fires where the ore burned could reach very high temperatures as the ore became white hot.

In a small mine once we had a fire in which the ore did not catch fire, but became hot enough to start timber burning and smoldering, creating large volumes of carbon monoxide which forced the evacuation of all working places. In this case there were no tailings available, as there was no concentrator. After isolating the burning area and shutting off all entrances for air, twenty-plus million gallons of water were pumped on top of the main ore body and allowed to filter down and cool off the burning area. This proved effective.

However, soon after the pumping started it was discovered

that a valve on a drainpipe through a concrete bulkhead on the bottom level had been left closed. As this drainpipe led to a fractured area of rock which normally drained the mine without need of any pumps, we were in danger of flooding the bottom level and the bottom of the shaft, as our small emergency sump pump was incapable of handling the quantities of water entering the mine.

When we arrived at the bottom level, there was already six feet of water and it was slowly rising. The valve was several hundred feet in from the shaft, and to try and swim that distance in the icy cold water at an elevation of over 14,000 feet above sea level didn't seem very promising, so we built a raft.

By the time it was ready there was only room between the water and the back of the tunnel for the raft and one passenger lying prone. The supervisor who had left the valve shut volunteered to make the boat trip and open the drain valve. By pushing himself along, using the roof of the tunnel, he eventually arrived at the concrete bulkhead, but by the time he got there it became evident that the rising water would not allow him room below the roof of the tunnel to return the same way. The only answer was to dive down to the closed 6-inch valve and open it, and so stop the water rising any further. This, fortunately, he managed to do and escaped with nothing worse than a bad head cold.

After one fire had been brought under control, a visitor from management in New York wanted to check the probable source of the fire. As the area contained no oxygen, but was only a short distance from a fresh air base, it was decided we could use the lighter chemical mask. Four of us, including the "top brass" visitor, set out and on arriving at the last fresh air base, proceeded to get our canisters working by gulping the air outside the mask and then breathing it through the system. The visitor got his mask working quickly, and strictly against safety rules and common sense, took off into the lethal area where the fire had started, by himself.

I had my mask working about 50% when I heard a shout

from inside, and glancing at my two companions I realized they were even less ready than I was. So fearing that our visitor was in trouble, and realizing that he couldn't be very far into the danger area, I quickly put on my mask and went after him. After proceeding about thirty feet or so, I saw I was going to have to get out, moving fast had already depleted my oxygen supply, and I was having trouble breathing. Also, as I had not had time to apply anti-fog spray on my glass face piece, I was having a problem seeing, and the residual smoke in the area didn't help. As I started to turn, a piece of timber tripped me and I fell heavily. Not bothering to get back on my feet, and virtually blind, I headed out on my hands and knees for the fresh air base as fast as I could go. At last, with no oxygen left at all, but not being able to see where I was, I had to take a risk and tear off my face mask.

I had gone right past my two companions at a gallop on my hands and knees, and they had merely noted my passing and continued to methodically prepare their equipment without saying or doing anything! As I pointed out, I didn't usually go around the mine workings in this manner, and they could at least have said "Good Morning" or something to let me know I was back to a safe area.

It turned out that our visitor was alright, and had only shouted for us to "Come on!" but the mask had muffled his cry so that it was unintelligible. However, even top management personnel should obey the safety rules.

In our rescue group we had one very large Canadian who always volunteered to be the "injured" person and be rescued. This meant that he didn't have to put on his breathing apparatus and merely got carried out of the mine by the rest of us. He was by far the heaviest member of the team, and transporting him through the old, semi-caved workings was no easy matter, so we decided to teach him a lesson. On the next practice session he volunteered as usual and was promptly told to wait at a designated point. As soon as we were equipped, we picked up the basket stretcher and set out to find our injured man.

The Canadian was lying on the floor giving vent to loud groans, he always had to ham it up, so we picked him up and strapped him securely in the basket so that he couldn't move. Once he was well and truly trussed, we took him a short distance to where a number of large icicles were hanging from the roof. These we broke off and packed around the helpless body on the stretcher. Although he was helpless, he had plenty to say, and was sputtering with rage as we carried out the next step in the plan.

Our underground bathrooms were pretty crude, and usually consisted of some old dynamite boxes, which when full were transported out of the mine and thrown away. For some reason these were known as "honey boxes." Not too far from the icicle location we had deposited some very ripe half-full honey boxes and here we transported our victim. We took a prolonged rest at this point, our self-contained oxygen breathing apparatus making us completely oblivious to the reeking atmosphere.

Not so for the luckless Canadian, the head of the stretcher had been carefully placed between the two most potent honey boxes. If the language had been bad before, it now reached fascinating new levels of invention, even as miners we had never head such a flow of expletives, and we stood in an awed circle, spellbound. Finally, when he became short of breath and started to turn blue in the face, we carried him out and released him. I never heard him volunteer again. He even turned down the role of injured man when it was offered to him.

Spontaneous heating of the ore would occur as the sulfide minerals in the core started to oxidize. In one of the mines they had areas where this happened very rapidly, a blast could not be left more than a few hours or the ore started to burn. When the sulfide ore burnt, it gave off great quantities of sulfur dioxide, a very pungent and acrid gas which made its presence known before it became lethal. In some areas though, the hot ore would start charring the support timber and if there was insufficient oxygen present, large quantities of carbon monoxide gas would be produced.

Carbon monoxide has no smell and gives no warning of its presence, and if it is suspected in an area, the only safe way to check it out is with a special canaries chemical detector. Sometimes in an area where stagnant water had been disturbed, hydrogen sulfide gas would be produced. This is a highly poisonous gas, but usually its smell of rotten eggs gives warning of its presence before any harm is done. However, prolonged exposure will reduce one's ability to smell the gas, and this can be serious.

It is not unusual to come upon areas of dead air in old abandoned workings, where the atmosphere consists principally of nitrogen and/or carbon dioxide, the oxygen content being too low to support life. We always tried to keep abandoned areas barricaded so that men would not enter them and run the risk of being gassed. At least once a year, in one or other of the mines, some miner would decide to save time and go to the bathroom in an abandoned area rather than walk to the designated location, and very often by the time he was found it would be too late to do anything for him.

Methane gas is usually associated with coal mines, and this gas or finely divided coal dust has been responsible for many explosions. In the Little Long Lac area of Canada it also occurs in the gold mines in dangerous quantities. Being light, it tends to collect in the backs of tunnels and drifts, raises being particularly dangerous. (A raise being a tunnel driven vertically or nearly vertical upwards.)

I worked in one raise that was 300 feet up, and due to connect to the upper level at the 400 foot mark, which had two separate methane explosions killing and injuring the members of separate crews. By the time I worked there we had a drill hole connection to the upper level which helped to clear the gas. Even so, the readings on the detector would sometimes go beyond the danger point and my Russian partner and I would leave in a hurry. The methane in these gold mines came underground in the water which seeped into the mine from the murky swamps on the surface. In places the water would froth like beer.

Apart from gas and fire, another danger would be water inrushes and occasional mud runs. Except for mines with natural drainage through rock fractures or low level tunnels leading out to surface, the pump room was of prime importance. In a big mine there would be more than one.

In one mine where they suffered from large volumes of acid water, it was Saturday night and a party was going on, when the mine captain received a phone call from a section foreman that the pumps had failed. On being informed that the situation was not too bad at the moment, the mine captain went direct to the mine office and simply pulled on a pair of long rubber boots over his suit trousers, and threw an oilskin on without bothering to remove his suit jacket. He and the cage tender slowly descended the shaft using the emergency signalling system for control, as they knew that the water had filled the bottom of the shaft, and was already up to the bottom level.

As the level station slowly came into view they observed the section foreman waiting for them with the water gently lapping around his ankles. The cage tender stopped the cage at the surface level of the water, and the mine captain stepped off the cage and disappeared from sight. He came sputtering to the surface and found his feet, and the water was up to his armpits. The section foreman was standing on the top of a large electric mine locomotive in order to keep dry. As this was strong acid water, after about a week the mine captain had little left of his suit except the buttons and zipper.

Time for a dance.

Chapter Nine
Fiesta in Santa Barbara

Santa Barbara, the patron saint of miners and artillerymen, also happened to be the name of one of the villages near the camp. Every year on her saint's day the villagers held a big fiesta, and all our miners made a point of attending. Fiestas in this part of the world last several days, so we virtually shut down for almost a week for lack of manpower. This never failed to infuriate the mine manager, who could never give a satisfactory answer to the head office to account for our loss of production. To New York a holiday is twenty-four hours off work, no more and no less; and yet, every year, in spite of threats and promises of bonuses and overtime, we invariably lost almost one whole week's production.

The Santa Barbara fiesta, I was intrigued to hear, always included a bullfight as the main and starting feature to get the whole thing off to a rousing start. I had never seen a bullfight, but had read *Blood and Sand* and *Death in the Afternoon* in the camp library, and was eager to see the real thing.

On the great day, two other gringos and myself piled into a half-ton pickup truck, well loaded with beer and pisco, and set off for Santa Barbara. We slowly descended from the barren altiplano of the mine area to the river valley, and gradually the lichens and tufted grasses of 13,000 feet gave way to small fields of potatoes and beans surrounded by crude walls built of rocks to keep the cows and sheep out. The sky was bright cobalt blue and cloudless, the sun hot, but because of the altitude, the temperature dropped abruptly when one entered the shade. A breeze blew up the valley and dispersed our dust cloud behind us as we bounced around the curves of the uneven road. As we wound down the river valley, stunted quinua trees, tall ragged Eucalyptus and cactus started to line the route. Wild geraniums and nasturtiums made splashes of color along the roadside.

As we approached Santa Barbara the valley broadened and the fields increased in size and the crops changed to corn and wheat. The river, an aquamarine color in the bright sunshine, kept us company down into the valley.

After over an hour's drive to cover the fifteen miles of rough road, we finally arrived in Santa Barbara. There were about forty adobe houses crowned by thatched or rusting corrugated iron roofs, tightly packed along the river bank, and the tower of a church could be seen rising from the central plaza.

We were greeted by the usual stray dogs and small children as we parked on the outskirts of the village. There must have been two hundred-odd locals present, most of them well primed with caña, beer and pisco. As the fiesta had only started the night before, money was still abundant and credit good, so everyone seemed to be well provided with bottles.

The main Plaza de Armas had been turned into a bullring by blocking the entry streets with buses belonging to the local cooperativa. The buses sported the bus line colors of yellow, mauve and green, and certainly added a touch of color to the festive scene. As is customary in these parts, the drivers had christened their vehicles and painted their names on them. We had "Bomba Atomica," "Una Más Y Me Voy," and "El Baron Rojo," present and accounted for. "El Rey de las Curvas" was still under repair after missing a corner and ending up in the river.

The women had on their best embroidered sleeveless velvet jackets over cotton blouses, full calf-length skirts covering many petticoats, and white straw hats, broadbrimmed and high crowned with wide black bands. The men all wore their best dark suits. Eager spectators were crowded on the sagging balconies overlooking the plaza and also on the tops of buses. After a lot of pulling and pushing we managed to join a group of Indian miners on top of "El Baron Rojo."

The priest was on the steps of his church which formed one side of the plaza, and surveyed the scene without enthusiasm, no doubt calculating the amount of overtime he would have to do in the confessional once the merrymaking was over.

There was a small monument surrounded by a flowerbed containing geraniums in the center of the plaza which appeared to me to be a hazard for the bullfighter. Noting my concern, my Irish companion, who had been to previous fiestas here, reassured me. He pointed out that it lent an aesthetic touch to the scene and should present no problem to a seasoned "torero," and added that the doves pecking away in the plaza likewise should not distract him from the moment of truth.

Someone gave me a handbill and I saw that it announced in large letters that Don Jesús Garcia Beltran, the bullfighter, was "Importado." I scanned further down the page and noted that he had been imported from the next village five km. down the valley. After recovering from this letdown, I was surprised to note that the bull had been "lent" by one Don Guillermo Rodrigues Marquez. I inquired of my neighbors, if instead of "lent" didn't they mean bred or donated, as after the fight there would only be a carcass to recover. With much shaking of heads and gesticulating of hands and bottles, they explained that the bull would not be killed.

After all, it was a poor village, and to kill a bull every year at the fiesta would be extremely stupid. Soon there would be no bulls left, and what then? No, señor, no wasteful bull killing for them, instead they used the same bull for several years running. Thus, after a few years the bull became an experienced fighting bull. Today's bull, for example, had fought in eight corridas over the last ten years, and this magnificent beast had now reached his peak of physical fitness as a trained fighting machine. I began to feel sorry for the bullfighter, and more than happy with my perch on top of "El Baron Rojo," well out of the way of this awesome brute.

The three-man orchestra was now belting out some lively Peruvian huaynos in place of the more usual paso dobles associated with a plaza de acho. Unfortunately, one quarter of the orchestra, the trumpet player, was missing as he had suffered an accident in the early hours of the morning. In the dim light of a local bistro, he had mistakenly picked up another man's full bottle of pisco, instead of his own nearly empty one. Apparently, nurses in the camp infirmary were still attending to him.

By now, after several slugs of the local white lightening, we were all getting into the spirit of things and were waiting impatiently for the spectacle to start so that we could give the local talent a big hand. Just as patience was reaching breaking point, Don Jesús, or "El Importado," entered the ring accompanied by a chorus of "Olés" and much handclapping. Our hero was a nervous looking little man, he kept licking his lips, and darting glances from one side to the other, as though the bull would leap out of the crowd at any minute. He was attired in a homemade bullfighter's costume of purple and yellow, and sported a miner's hard hat securely fastened under his chin with a piece of string, on his feet he wore football shoes.

I noted the hard hat, and was glad to see that our safety engineer was getting his ideas across to the local workers. The football boots no doubt were to help him maintain his footing while executing some of the more difficult passes with his cape. By this time I was beginning to wonder what kind of a fight I would see, and, frankly, was resigned to seeing a pretty tame exhibition. Little did I know what an excellent fight would crown the day.

One road opposite the church had been blocked off by a barricade with a gate on it, and this was the bull's entry way into the ring. At long last, accompanied by much cheering, the gate was opened with a scattering of doves and feathers, and our fighting machine entered the ring. Shouting and shoving, the five small boys who pushed the bull into the ring managed to head him towards "El Importado" at a shambling trot.

For one brief instant Don Jesús held his ground, and then with no further hesitation, turned and shot across the ring like a bullet, straight into the church, his fighting cape left on the ground of the ring. Instead of following him, the bull started to trundle round the plaza trying to find a way out.

I thought at the time that unless they could pour a considerable quantity of alcohol into "El Importado" to encourage him to come out of the church, that the fight would be over, but not so. Drunks now started to enter the ring waving their coats or capes. The bull seemed mildly interested and made a few half-hearted passes with his horns as he trotted past the would-be toreros, obviously he was still intent on finding the way home. Meanwhile, true to the traditions of Madrid and Mexico City, the guardia civil (police) entered the ring, seeing their friends in the ring being beset by the local gendarmes, other drunks piled into the plaza to join the fray. In no time at all there were a dozen lively fights going on, police vs. the citizens. A large young miner in front of us was doing beautifully, when a bottle presumably aimed at one of the police, missed its mark, hit him behind the ear, and put him out of the rest of the action.

Meanwhile, the bull continued to trot around the plaza, impartially prodding all who got in his way and leaving a trail of destruction behind him. He had by now collected a coat on one horn, and the drunken owner, firmly attached to the bull's tail, was being dragged around the plaza shouting Quechua oaths as he tried to make the bull stop so that he could regain his jacket.

The grand finale arrived when the municipalidad balcony collapsed. Luckily, this sagging structure, which had probably been built in the time of Pizarro, was not too high above street level, and it let go fairly slowly, spilling the village mayor and dignitaries into the path of the circling bull. In short order, a dozen newcomers rushed up the church steps seeking sanctuary.

The angry village elders, suffering more from a loss of "dignidad" rather than any physical hurt, proceeded to blame "El Importado" for the whole fiasco. As they worked themselves up into a fury, they started to shout and manhandle the bullfighter. The priest, a fiery-tempered little Spaniard, fruitlessly tried to quiet them and make them respect the sanctity of his church. Finally losing patience, he carefully removed and pocketed his glasses, and with a beautiful right, coldcocked the mayor right on the church steps.

At this point, just when a full-scale riot appeared to be starting, the women decided to take a hand. No doubt to save the fiesta, wives and girlfriends of the combatants entered the plaza in a swarm. Hardened by long hours of work in the fields, they were a force to be reckoned with, and as each one grabbed her man and dragged him off, the spirit went out of the fighters and they offered no resistance. The orchestra struck up a rousing huayno on cue, and soon the couples started to dance.

As peace and good cheer returned to the fiesta, the victorious bull was led by the boys from the plaza where he'd started to munch the geraniums, doves scratching around his hoves. I often thought afterwards that this was a much nicer ending to a bullfight than the real thing.

Chapter Ten

Challenging Situations

Apart from normal production problems in the mines, there were always other difficulties to be resolved. In Canada, it was not uncommon for a mining camp to be threatened by a forest fire, and more than one mine has lost its surface buildings and concentrator due to these fires. The forest rangers could conscript anyone they needed in order to fight fires, and if they thought it necessary, would come in and take the whole mine crew. This happened on one occasion in the Selkirk Mountains of British Columbia. Much to the miners' disgust, we had to walk eleven miles into the fire which was burning high up on the side of a mountain. We were immediately put to work on cutting a fire lane, or long open area, below the base of the fire. This was to stop burning debris from falling downhill and spreading the fire in that direction.

We soon discovered that frequently a burning tree would fall with a loud crash, and then go tobogganing downhill like a great big flaming torpedo right through the firefighting lines, and if we were not careful would start a fire below and behind us. The hillside was covered with a thick coating of pine needles, and the fire would creep under the blanket of fresh needles. The only evidence would be the heat rising from the forest floor. Every now and then a pine tree would get "cooked" by this creeping fire to the point where its gums and resins would start giving off inflammable gas, then suddenly with no warning the tree would explode in a sheet of flame.

The hillside was so steep that the only way to sleep was to drive two pieces of wood into the ground, then lay a horizontal pole across these on the uphill side. With this to support our feet, we no longer slid downhill in our sleep. I say "our" because there were not enough blankets, so we slept four or five together. We had one very short Canadian with us who made the mistake of sleeping in the middle the first night. We were all filthy dirty and smelled to high heaven, and with all our feet resting on the same pole, the short Canadian's head kept disappearing under the blankets. All through the night Frenchie swore as he tried to get his head out into the fresh air.

Food was short because a lot of cases that were dropped to us by plane had fallen into a swamp and disappeared from sight. Finally, one of the firefighting crew was killed by a falling branch or "widow maker," and we had the task of carrying the body out through the almost trackless valley. We were very glad to get back to camp and go to work in the mine again.

In Peru, it was company policy to help out the local communities, and a budget was set up for this purpose. We always insisted that whoever received any help should also chip in and do their share of the work. The company would provide some equipment and the village council would provide the manual labor.

Usually this work would involve the construction of new roads or improvement of existing roads, and sometimes help with a water supply or bridge. One superintendent, having agreed to help a local community to widen their only access to the outside world, invited the village elders to view the first blast. There was a thunderous roar and a cloud of smoke, and to his horror, the whole road disappeared off the side of the mountain into the canyon below. To his credit, he rectified the error, but the work took a lot longer and cost a lot more than the first estimates.

One small town located along the banks of a fair-sized river in a steep valley sent us a desperate call for help one Sunday evening — it always happens on a Sunday. It had been raining very hard for several days and one of the side valleys that came down through the town had had a flash flood which had caused some damage. The flood had also brought down large quantities of mud and rocks which had dammed the main river, causing it to rise, and the rising waters were gradually undermining the adobe houses and causing them to collapse.

Two of us arrived in the dark, in a pouring rainstorm, with buildings crashing into the main street, to see what could be done. We called back to camp by radio and arranged for some earth-moving equipment to be moved down to try and clear the side valleys which were pouring water into the town as their normal flows were choked off. The next day we got a better look at the situation and realized that to make any real headway we had to clear the dam blocking the main river, but the water was too deep for the equipment.

Necessity being the mother of invention, we built a raft and passed a rope across the river and secured it to trees on each bank. Then our blasting engineer, armed with large shopping baskets obtained from the villagers, loaded the baskets with dynamite and weighted them with rocks, and then lowered them underwater along the length of the dam. In doing this he managed to fall in twice, but luckily kept a hold of the rope, and each time managed to regain the raft which was also attached to the rope.

The blast was quite spectacular, as a huge sheet of water shot into the air, and also successful, as it broke up the obstruction enough, particularly in the center, that the river started to drop and we could get our tractors to work. Meanwhile we had organized food and clothing for the people evacuated from the danger area and in conjunction with a small group of soldiers, were knocking over buildings that threatened to collapse. Each evening after work we would go down to inspect and direct the cleanup, eventually getting home about midnight. In all, some three villagers were lost and fifty houses destroyed. The mayor later thanked us for our help.

We had a steep valley where we used to deposit our waste

rock from the open pit. As no one lived in the valley close to where we were dumping, I am afraid to say no one took much notice of the way the waste dump was being formed. Early on a Sunday morning I got an excited call from a mine owner who operated a small concentrator about three miles below our waste dump. He didn't seem to be making much sense as he kept saying that our waste dump had caused a landslide which had damaged his mill. At first I thought he must be drunk, as in no way could our modest pile of waste rock go three miles down the valley and interfere with his operations. Anyway, being awake now, I decided to investigate, and what a sight!

Some 50,000 tons of rock had fallen off the face of our waste stockpile and hit the bottom of the valley with a jolt. The valley bottom consisted of an ordinary-looking green meadowland, some swampy areas and a small stream meandering down the center. What we hadn't realized was that all this ground was super-saturated with water, and that the jolt of the falling rock had been enough to liquefy the whole valley bottom. Thousands of tons of liquid mud were slowly pouring down the valley, like a great black glacier, up to twelve feet deep and at the rate of about thirty feet per minute. It stretched for about four miles down the valley and was still moving.

The small concentrator and part of the camp were well and truly buried. Luckily, no one was missing, but people were already clamoring to be paid for missing livestock, both real and imaginary. Tractors and front-end loaders were brought down from the mine to try and clear the road, concentrator and camp. As the operators inadvertently went off the road or into a deep patch of mud, the equipment would stall, and the operators would have to be rescued by ropes. In a very short time the scene resembled the retreat from the Chosin reservoirs [Korean War] with equipment stuck in the mud and pointing at odd angles all down the road.

There was no way of stopping this black flood, and as the valley was more open further down, we just pushed it on downstream and dispersed. Finally, when it dried, the solid content of the mud shrank to only one or two feet. We fixed up the small concentrator and camp, but had to pay for a quantity of concentrate which the owner said had been washed away by the mud, and which was the equivalent of about six months of normal production. We also paid for numerous sheep, chickens and a horse.

These mud flows are not uncommon in the Sierra and sometimes a seismic shock or just the amount of water reaching a certain percentage will start them moving, and there isn't much you can do about it except get out of the way.

Snow avalanches are a different story, and something to be treated with great deal of respect. At Cerro's Rio Blanco mine in Chile (nationalized by the Allende government in July 1971) we frequently used to get a total snowfall during the year of over sixty feet. As the road up to the mine led through a narrow and very steep glaciated valley, avalanches were a serious problem during the winter season. We used a 75 mm. recoilless rifle and special mortars operated by compressed air to shoot explosive projectiles at the accumulations of snow overhanging the roads in an effort to remove the danger. All traffic would be stopped during the day. Convoys of men and materials could only pass between the main camp and the mine and concentrator in the early hours of the morning.

Temperature changes very often sparked the movement of the snow, and as a shadow crossed a mountain slope there would be a crack like a rifle shot and an avalanche would come streaming down the hillside, shortly to be followed by another as the shadow moved along the snowfield. All buildings in the avalanche areas had to be carefully located and specially constructed of reinforced concrete. The concentrator itself was built completely underground and both tailings and concentrates had to be taken down the valley in pipelines, the concentrate pipeline being some twenty kilometers long.

The key to the whole operation at Rio Blanco was recognition of the snow and avalanche problem, and all the original planning for this mine was oriented to overcome winter operating conditions. The three main buildings were in avalanche-free zones near the concentrator. The "Hilton," as its name suggests, purely for accommodation, and "Rock Point," the mine office, were connected to the working areas by underground tunnels. Crews established in these buildings before storms occurred could go to and from the working places without going outside.

The third building, "Lagunitas," was the headquarters for the Snow Department, which had complete authority over all surface movement, and reported directly to management during the winter months. The Snow Department also had four refuges spaced along the canyon to house men and equipment for clearing the highway, and to serve as shelters for anyone who got cut off by slides while traveling up or down the road.

In order to ensure continuous operation, the mine crews were split into three groups, each group working twelve days with two days off, the work periods overlapping. Thus there were always two production groups in residence in the work area.

At the Cerro de Pasco mine in Peru, the company's first mine for which it was named, we had an open pit that was gradually encroaching into the local town. A large group of engineers was employed buying up properties and also building a completely new town to replace the old one. The layout and construction of the new town was coordinated with various government planning agencies and was built with running water and inside toilets for all the buildings, plus proper draining and sewage facilities for the whole complex, things that were conspicuously absent in the old town. In spite of everything that was done to try and make the transition from the old town to the new one as smooth as possible, problems between the company and the townspeople were not uncommon.

The biggest complaint was that blasting in the open pit cracked the walls of their houses and damaged their personal belongings. A most stringent control on blasting was maintained and records using seismographs were kept to make sure that there would be no blast damage. Despite our efforts, we would invariably receive a list of claims after each blast. On occasions when only secondary blasting had been carried out, that is blasting of large boulders too big to be handled by the equipment, we would still get claims.

This was because the secondary blasting caused more noise than the big ore-breaking primary blasts, and damage seemed to be associated with noise rather than ground vibration, the ground vibration from a secondary blast being virtually nil. A typical case would be the man who produced a clock that had stopped at twelve o'clock, which he claimed one of our blasts had knocked off his table and stopped the mechanism. The fact that he claimed it had kept perfect time before falling off the table, and that we always blasted at 5:00 p.m., did not stop him from pressing his claim!

Occasionally, one of the town lawyers would palace a "denuncia" against one of the staff, and the civil authorities would issue an arrest warrant. Being on good terms with the guardia civil, they would let us know that they had orders to pick up someone at a certain hour so that he had time to leave town. In one instance which involved the plain clothes branch, and not the guardia civil, the message didn't get through in time and I had to escape through the back window of the office and leave town hiding on the floor of a pickup truck.

Hiding out in Lima or one of the other mines after making one of these escapes, word would come through that the legal process had proceeded to the point whereby one had to return to the district. This would mean sneaking back to the mine in the early hours and getting admitted to the camp hospital. The authorities would then be informed that you were back within their jurisdiction, but were sick in the hospital. The guards would check at shift change to make sure you were still there.

The doctors would select some mild ailment for the record,

as the town or police doctor would sometimes be sent around to have a look at you. One time I had been given typhoid as my ailment, when much to the doctor's surprise, he found from his routine checks that I really did have a mild case.

On one occasion, the superintendent of the pit was hauled into court and accused of murder. An old lady had died of a heart attack some hours after a blast, and the relatives were looking for compensation. The company lawyer obtained the superintendent's discharge after several court sessions, but it was often an uncomfortable situation. The mining operation was big enough and complicated enough without such added harassment.

The day came when we had to knock down a church which bordered the edge of the pit. Not just one, but two replacements had already been built and accepted by the local bishop and authorities, who had given us permission to proceed with the demolition of the old church. However, factions in the town swore that they would never allow us to lay a finger on that church, and they would defend it with their lives. One night at midnight, and surrounded by armed police, we went to work and tore it down. I'll always remember meeting a good friend, a Peruvian engineer, coming out of the church at 2:00 a.m. with the bell in his arms (it was being saved for one of the new churches), asking me to please be sure to not let his mother know he had helped to knock over a church! After this event there was no more church trouble, the opposition decided to think of other things instead.

We also ran into a lot of problems trying to cut one of our roads which bordered the pit. As it was a quick route from one side of the work area to the other, the townspeople objected strongly to its removal. It was on our own property and we had constructed it in the first place, so we did not see eye to eye with them on this.

At midnight one night another engineer and myself decided to go ahead and cut the road anyway, and present everyone with a "fait accompli." We took a Caterpillar D9G tractor for the job — nothing was too good for the job. The tractor was equipped with rippers, and so we went to work to tear out the road. The first pass opened up a narrow opening to some old workings. The next pass cut the main town sewer line which had not been marked on any of our plans as being in this area at all. Ankle deep in this vile smelling flood we stopped our project to discuss what to do next, when we heard faint cries for help further down the road.

On investigating, we found that an innocent passerby had fallen about fifteen feet into the old mine workings and the sewage was cascading down the hole on top of him. Quickly improvising a rope out of raincoats, we managed to haul him free of any injury, but smelling worse than a skunk. At this point we decided to call off the operation and take our smelly friend over the to the mine showers where he could get cleaned up. We also gave him enough money to keep him eating and drinking for several days as it was a bitterly cold night and in his soaked-through state he was shivering violently.

A sequel to this episode was the complaint next day from a small concentrator nearby that we had cut off their water supply. Apparently for some years they had been using the town sewage for their small flotation plant. With no problem we quickly hooked them into our drinking water mains, which we were sure they would much prefer to the stinking supply they had used previously. Not so, within a few days we got an energetic complaint that their recovery had gone to hell, and that on top of this, their costs had gone up as they now had to use chemical reagents, whereas before the unadulterated sewage water had done a magnificent job without the aid of reagents!

Every few years or so, usually during wage negotiations, one camp or other would have a serious riot. After the dust had settled and a few agitators had been fired, everything would go back to normal for another three or four years. These riots were pretty nasty, and often ended up with the police shooting someone. Police were present in all the camps, but not enough to handle a bad riot even though reinforcements would usually be sent in when trouble was expected.

One case is enough to describe what could happen. All the mines had been on strike for several days and relations all over between the union leaders and the company had deteriorated badly. In one of the most isolated camps, the union leader had managed to jump one of the local policemen and had taken away his submachine gun. During the struggle the gun had gone off and wounded a young boy in the leg. There were not enough police present, nor did they have anyone there of sufficient authority to try and get the weapon back. The union leader either realized he had committed a crime which would land him in very bad trouble or else he let the possession of the weapon go to his head, probably a bit of both.

He then took a crowd of strikers to the staff houses, which were a good distance away, and started throwing rocks at the houses and cars. The mine superintendent ventured out of his house to try and calm them and persuade them to return to the camp. On going outside, he was knocked down and taken hostage. Seeing him knocked down, two other staff members went out to help and they were promptly taken prisoner along with him. The crowd then returned to their camp with their three hostages who'd had their arms tied behind them with wire. Several of the strikers were armed with chains, and the mine superintendent was repeatedly knocked to the ground during the trip. By the time they arrived at the union headquarters, the superintendent was in very bad shape and hardly able to stand.

After much arguing, the two staff members who had also been taken prisoner managed to persuade their guards to bring the doctor to attend to the superintendent's injuries. After the doctor had done the best he could, he insisted that both the superintendent and the boy with the leg wound must be sent to a proper hospital, with dire warnings that the former might die of his injuries. The doctor finally had his way, and the two injured left that night in the ambulance.

The two remaining hostages then spent several most unpleasant days under guard at the union headquarters. Once or twice a day there would be meetings to decide whether the prisoners should be killed immediately if a rescue was attempted, the arguments pro and con and subsequent voting all clearly audible to the captives.

After some days of imprisonment, a truck drove up and parked in front of the union hall. One man in worker's clothes jumped down and went straight inside to the hostages and told them to quickly lie down on the floor with him. No sooner had they done this than a burst of gunfire came through the thin corrugated iron walls of the hall from the disguised police outside. They stormed the building. The union leader had dropped in the first hail of gunfire through the walls. A dynamite bomb thrown by a striker luckily missed the truck and exploded harmlessly down the hillside, however several men were wounded before the battle was over. Some days later a battered and permanently scarred mine superintendent returned to take charge of his mine.

This instance was perhaps more dramatic than most, but shooting and bloodshed were not uncommon. When a crowd of Indians is aroused it can be very unpleasant, and they say the only thing worse is when their women go on the warpath. There was a story that years ago in a nearby village a crowd of women angered over no one remembers what, attacked and killed the local prefecto, and then dragged his mutilated body around the plaza in triumph.

Cerro de Pasco mine work force with Brian as Mine Superintendent, Asst. Superintendent Mike Dzierla, to his right.

Chapter Eleven
Mines and Mines

Coal mines are something else. As the seams are usually fairly flat and the thickness is often less than the height of a man, they can be extremely uncomfortable to work in or visit. Spending several hours crouched, or on one's hands and knees, brings on aches and pains in muscles one never before knew about. Besides the discomfort, the mines are naturally dirty with fine coal dust which gets into the eyes, ears and nose, and which can also make an explosive atmosphere. Methane gas is usually associated with coal mine explosions, but very often they are caused by fine coal dust suspended in the air.

As the seams are mined out, the ground is encouraged to cave close up to the opening that is made. This takes ground weight off the working face, and makes it safer for the working crews. However, it puts great pressure on the tunnels giving access to the working places. If constant repair is not carried out, the supporting steel in these tunnels is bent and crushed out of shape.

In France, to take a short cut from one area to another we would frequently lie flat and then pull ourselves a couple of hundred feet with our elbows under the distorted steel and rock. To meet rats coming the other way was a frequent occurrence, you would see their red eyes reflecting the rays from your lamp. After a few annoyed squeaks they would move out of the way and let you through. It was not uncommon to be halfway along one of these openings and then hear the ground move. The first few times it feels enough to turn your hair white and then you get used to it.

These French mines had rats underground because they used many horses in the mine to haul cars of coal to the shaft. Horses can be just as bone-headed as any old miner. We had one that would make no more than fifteen trips before and then fifteen after lunch. Any attempt to up production and old "Gably" would deliberately derail the cars on the sixteenth trip. If he felt particularly ornery, he might do it more than once. I never heard of anyone trying to make a seventeenth trip, either before or after lunch.

Not all the coal was hauled by horses, we also had conveyor belts in some areas. These were very convenient to ride on as they quickly got you to your destination, however, it was strictly forbidden in the mine safety regulations.

One time we had a visit by some eminent engineers, and they went underground with the mine manager and assistant manager to see some new coal cutting equipment. When it came time to leave, the manager, leading the way, jumped on a moving belt and was quickly followed by his assistant and the visitors. As they neared the end of the belt run, the manager rose to his hands and knees to be ready to jump off. Just before he was able to jump, a protruding nail hooked into his belt and held him fast, his only recourse was to try and run backwards on his hands and knees as the conveyor was still moving forward. The assistant, who was next in line, had no time to jump clear as before he realized what was happening, he crashed into the manager's rear end.

The shock of the collision broke the belt free from the nail, and both men, somewhat tangled, sprawled forward on the moving conveyor. Before they could sort themselves out the belt arrived at the end roller and deposited them both in the bin in a flurry of arms and legs. Luckily, neither man suffered any injury except to his dignity. The choleric manager, being the senior, took out his rage on his assistant, much to the amusement of the visitors.

The manager was a vindictive man and planned a quick and

effective revenge on the engineers who had enjoyed his predicament. After a long and very rich French lunch complete with wines and cognac, he arranged a visit to a new shaft that was being sunk nearby and was already down about six hundred feet.

He packed all the visitors into the sinking bucket, so tightly that they could hardly move, and then signalled the hoistman to lower away slowly. It is customary when shaft sinking to use a cable that won't spin, as near the bottom of a shaft being sunk there is nothing to stop the bucket from spinning. "Guides" cannot be carried closer than fifty or sixty feet to the bottom or they will be damaged by the blasting. For some reason this shaft did not have a non-spin cable, possibly shortage of supply resulting from the war.

Once the manager knew the sinking bucket was near the bottom of the shaft and free of any restraint, he signalled the hoistman to stop lowering. During the few minutes the bucket remained suspended it revolved rapidly, first in one direction and then the other. Several of the visitors with weak stomachs were no longer able to hold down their lunches, and being packed in so tightly, no one could move out of the way. When they eventually arrived at the shaft bottom and were helped out of the bucket by the grinning shaft sinkers, they were a very sorry looking crew, all so dizzy they could hardly stand.

In order to keep the coal mines clear of explosive methane, large quantities of air are continually being blown all through the mine to remove and dilute the gas emanating from the coal seam. This air is split and directed into different working places by using control doors and auxiliary fans. In a mine with many miles of openings it is a complicated undertaking made more difficult as there is constant change in the working places, connections and tunnel advances.

Lack of enough ventilation in an area will cause an explosion which will often produce coal dust which will itself explode in a secondary explosion. The pressures on the doors can be quite high and make them difficult to open so that air locks and double or even triple and quadruple door systems have to be used. Once the door is cracked open and the pressure equalizes on both sides there is no problem, but to make the initial crack often takes quite a lot of force.

One day underground with our German Mine Captain at Cerro's Goyllarisquisca (known by all as "Goyllar") coal mine, we were going through a series of these doors which were quite hard to open initially. Hans, leading the way, was showing off by charging the doors with his shoulder. At the third or fourth door, just as he launched himself wholeheartedly at the door, a miner opened it from the other side. Meeting no resistance, Hans shot forward through the open door and measured his length in three inches of water, mud and coal dust on the floor of the tunnel, his glasses catapulting off his nose into the muck. I had always heard that German was an expressive language, and I must admit as Hans groped for his glasses in the dirty water, he gave vent to a stream of language that certainly got the message across to the frightened Indian who unwittingly opened the door at the wrong moment.

As the high grade veins and ore bodies of the world are mined out, miners have been forced to mine large low grade deposits in their place. The only way to do this economically is to move and treat in concentrators very large daily tonnages. When the deposit is close to the surface this can be done by the open pit system which is presently responsible for most of the iron and copper ores produced in the world. This method utilizes large electric shovels loading into special off-highway dump trucks that can carry anywhere from 50 to 200 tons or more per trip.

There are a number of mines of this type, moving around 100,000 tons of ore per day. On top of this, they have to move a further 100,000 to 250,000 tons of waste per day in order to continue mining the ore at deeper horizons. The walls of a pit cannot usually be steepened much over 37.5° or there is danger of a wall collapsing and burying men and equipment under hundreds of thousands of tons of rock. In order not to interrupt the work of the loading and hauling, equipment blasts are not made every shift but perhaps once a week, and can involve several hundreds of thousands of pounds of explosives. This can cause problems if the mine is close to a town or village and great care must be taken to control the blast so that the shock wave will not cause any structural damage.

Sometimes these blasts will not break the rock into small enough fragments for the loading equipment and these large boulders then have to be broken by "secondary blasting" or by a crane with a drop ball. Secondary blasting can be accomplished by either drilling a hole into the boulder and filling it with explosive set off by a cap and fuse or by using a "plaster" or "bomb." The latter is an explosive charge laid on top of the boulder, with the explosion breaking the rock as though it were hit by a hammer. Both methods produce a lot of noise and the former tends to throw rocks in all directions and is difficult to control.

When blasting you can see the rocks arching up into the air and coming towards you like a shower of mortar bombs. This is no time to be caught without a hard hat, there is just no way two guys can get under one hard hat, I know, I've tried it. If it looks really rough, get under the nearest vehicle, better dirty clothes than a hole in the head.

The open pits are developed by mining a series of levels on "benches" that might be 30 to 50 feet high, depending on the size of the equipment in the mine. The benches are connected by inclined ramps, 8 to 10% gradients, so that the equipment can move from one bench to another. As the pit is deepened, the benches are pushed back in order to maintain a safe wall slope and this will mean moving waste rock.

When you have to move more waste rock per ton of ore than the recoverable metal in that ton will pay for, then you have reached the economic limit of the mine. On occasion, by changing equipment and lowering your cost per ton moved, the life of the mine can be extended and, of course, variations in metal prices are another important factor in deciding the economic life of a mine.

Open pit mining is the cheapest system of mining hard rock. Dredges and "monitoring" are cheaper, but can only be used in soft formations. Monitors are special nozzles which can direct large streams of water under high pressure at the material to be moved. They are normally used for gold or tin alluvial deposits where the washed material flows with the water through a gravity separation concentrating plant.

The old timers used "sluice boxes," these were long wooden boxes, open at the top, set on an incline with a series of riffles or baffles an inch or so high, fastened to the bottom of the box and at right angles to the flow. The heavy ores of tin and tungsten or gold nuggets travel along the bottom of the box and catch behind the riffles. Periodically the flow through the box is stopped, and the "pay dirt" is cleaned out from behind the riffles. For a very fine gold, rough jute cloth or corduroy can be used. The ancients used sheeps' fleeces, hence "Jason and the Golden Fleece," it wasn't golden but contained fine gold, as it had been used in the bottom of a sluice box.

When a large low grade ore body is too deep to make it economical to move the large tonnages of waste to get at the ore, or perhaps due to very heavy snowfalls an open pit is not feasible, then other systems have to be found. The first in low cost after open pitting is "block caving." In this system the ore body is planned out in blocks of say 200 feet by 200 feet. Some 200 or 300 feet below the top of the ore body a level is established. Three or four blocks are then semi-isolated from their surrounding rock by a series of tunnels, raises and slots.

The blocks are then undercut by removing a slice of ore some 12 to 20 feet high from underneath them. This is normally done by retreating towards a safe exit, and leaving a series of "draw points" between the undercut floor and the working level 20 feet below. As the ore breaks and crumbles under its own weight, it is pulled from these draw points to keep it moving. Obviously, you have to have an ore body with a fine amount of fracturing for this system to work correctly. If the ore body is not fractured enough, the rock will break into very large boulders and will block off the draw points, and it will be impossible to keep a steady flow of ore going to the concentrator.

Sometimes the ore will not cave spontaneously at first and has to be helped by long hole drilling and blasting or by increasing the undercut areas and the length of the spans of unsupported rock. On other occasions the ground may start caving satisfactorily, but after working up a certain distance it might arch and stop. This state of affairs can become extremely dangerous.

The men below go on merrily pulling their production quota from the draw points, meanwhile a large cavern starts to develop between the surface of the gradually descending broken ore and the roof on back which is no longer caving, but staying tight like a huge vaulted cathedral ceiling. When the back finally does come down, very likely a million tons or more in one helluva crash, the displaced air is forced into the workings with enough power to pick men up right off their feet. There have been times when men have been hurled over 100 feet along a tunnel and smashed against the rock and killed. The air blast can also tear out supporting timber and wreck equipment with its explosive force. Furthermore, if a sufficient cushion of broken rock has not been left above the working level, the sudden blow from the cave can damage or completely crush the working level below the undercut.

On one occasion a hung-up cave had a different result, fortunately with no fatalities. The cave must have temporarily halted very high up in this particular mine which had been working for many years. In the cavern that developed as the miners continued pulling ore, a large quantity of dead air with a very low oxygen content accumulated. When the back commenced caving again, with no warning to the crew down on the working level, it forced this dead air down through the broken ore and into the working places. Suddenly men started dropping like tenpins, they would stagger, gasping for a few feet and then collapse. Luckily, the new shift was just coming on and with the help of the forced ventilation from the mine's fans they managed to pull everyone clear and revive them.

If the rock is too hard or blocky to use a block caving system, then depending on the size and shape of the ore body, various systems using long drill holes and blasting are utilized. The "holes" left after the ore is withdrawn are sometimes filled with waste and sometimes the rock walls are allowed to cave and fill the openings. Large diesel equipment is used underground to load and haul this broken ore to the ore passes or shafts and from there it goes to the concentrator. This diesel equipment can operate on very steep grades and often the workings are connected to the surface by ramps, declines or inclines. In the Kuruna mine in northern Sweden, tourist buses make daily trips from the surface to the working areas, and a comely hostess gives an explanation of what is going on to the spectators.

The Swedes have some very up-to-date mines, and underground cafeterias with waitresses are not uncommon. Also in one or two mines small flower pots greet the visitor when he steps off the cage onto the level.

The most immaculate "mine" of all is run by Altas Copco, a large Swedish manufacturer of rock drills and other mining equipment. At their home office in Stockholm one steps into an elevator and is taken down to their demonstration mine cut out of granite right under the head office. A business suit and a hard hat is the normal attire. It was certainly a nice change from some of the dirty holes I have spent time in.

Chapter Twelve
Fishing Lore

The lake was called Huascracocha, and it lay in the central sierra at an elevation of just over 14,000 feet. It was about two miles long and a half a mile wide with no visible outlet and was surrounded by low grass-covered hills on three sides. On the east side where the water was deepest was a steep escarpment with many rockfalls and talus slopes that were home to numerous Vizcacha, the rock rabbit of the Andes. Although they are the size of a rabbit, there the resemblance ends. They have short ears and a long bushy tail like a squirrel, with thick grey fur coats.

The lake was home to countless duck and teal, and occasionally one would see some of the pink Andean flamingos wading cautiously in the shallows. We had fished the lake a few times, but no one had ever caught anything although we had often seen quite reasonable rainbows cruising offshore. There were a lot of Chalwa, small bony indigenous fish, which the imported trout enjoy as food. This probably indicated the trout were too well fed to be easily caught. Consequently, we usually went to more productive fishing holes.

At the time I'm speaking of, we had two visiting American engineers staying with us, and one of them was keen on hunting so I promised to lend him my 16 gauge and take him out on the following Sunday. As I only had one gun, I figured I would take my rod along and give the lake another try.

We drove out to the lake in a company half-ton pickup, leaving after an early breakfast. Although the sun was shining out of a cloudless sky, at that time of year one could often have snow and sleet storms in the afternoon, as it was the tail end of our rainy season. As we drove along the dusty road we passed small flocks of llamas and alpacas grazing close by the roadside. Momentarily they would stop grazing and raise their heads to give us supercilious looks before trotting off. Farther away flocks of sheep grazed on the hillsides looking like slowly moving maggots.

When we arrived at the lake and parked the truck, I gave my friend directions toward the most likely area to find "pato reales," the big brown sierra ducks with purple and white flashing feathers near the wing tips, and which are fair eating. I also warned him against shooting any Huachuas, these big black and white birds are a species of goose, but as tough as an old boot to eat. I then picked up my rod and gear and headed over to the deep east side to start fishing.

I had been fishing for maybe an hour with no sign of a fish, and stopped to sit and have a cigarette. I was not particularly disappointed as I had not really expected any action and was just passing the time while I waited for the bird hunter.

While I had been fishing, a little wizened old Indian who was looking after a small flock of alpacas drifted over to watch me. He was spinning raw wool into yarn with a hand distaff, and was dressed in coarse home-woven clothes with a knitted woolen hat with long ear flaps and a pair of homemade llama hide "yanquis" (sandals) on his feet. I wondered if he understood Spanish or only his native Quechua, so offered him a cigarette. The wrinkled face broke into a big gap-toothed smile as he accepted the cigarette and thanked me in Spanish. We then discussed the weather and the health of his alpacas before he asked me if I'd had any luck. I told him no, and then added defensively that there were very few fish in the lake, and that they were hard to catch.

He listened politely to my excuses, and then remarked, "Of course, of course, but the Señor is not fishing in the right place."

"Oh, and do you know of a better place?" I asked, thinking that I had nothing to lose if the old boy took me to some other part of the shoreline, and it just might prove to be better than where I was, it certainly couldn't be worse!

"Sí, señor, not only will I show you the place, but I will also show you the fish, and they are 'grandasos'!"

This sounded interesting, so I offered him another cigarette and we scrambled to our feet and I took off with the Indian in the lead. The old boy was no slouch, and in his element in this high thin air, and he soon had me panting to keep up with him. As we proceeded, I realized he was taking me round to the shallow side of the lake a few hundred feet from the road and my expectations diminished. I had never fished that side because I knew that the shore below the road was only eighteen inches to two feet deep for many yards out from the beach, and did not appear to be very promising for decent sized trout. However, it was on the way back to the truck so I didn't say anything about going on a wild goose chase.

We cut inland from the lake and climbed a short rise and then we arrived at the top and the Indian stopped and pointed. "There, Señor."

I was still puffing from my exertion and looking down all I could see at first was a shallow sandy bay on the other side of the promontory about sixty yards below us. Then my eyes came out like organ stops and I caught my breath. There were twenty-odd long grey shapes cruising around in the two feet of water, for all the world like a U-boat wolf pack.

The startled Indian must have thought I was calling on my God when I burst out with an explosive "Holy Cow!" I gave him a full pack of cigarettes, and told him to sit where he was so as not to scare the fish, and then carefully removed my boots and crawled toward the lake. I need not have worried, the fish were completely unconcerned and had obviously never been fished before. The only reason I didn't catch a fish in my first three casts was because they were all trying to take the lure at the same time and getting in each other's way.

On my fourth cast I played it smart and cast off to one side where a straggler had a good head start over the rest of the bunch and sure enough he took the flatfish at his first pass. I quickly worked him out of the bay and along the shore a ways so that we wouldn't disturb the other fish. Fifteen minutes or so later I had a nice eight-pound rainbow on the bank, and was feeling as though I had broken the bank at Monte Carlo.

I went back to the little bay, but not bothering to crouch or move stealthily this time, and sure enough, the fish continued with their aimless circling, taking no notice of my approach. As I prepared to cast again, I realized what must have been the reason for the concentration of trout in this particular area. I could hear running water and somewhere under my feet was the subterranean exit from the lake. This was the only difference between this little bay and all the others, and it was always here that the fish congregated.

When my hunter friend returned an hour later, I had three rainbows on the bank from seven up to nine and a half pounds, and had stopped fishing as I wanted to leave some fish for a future visit. My Indian guide had happily departed with most of my cigarettes and a nice cash "propina."

Two weeks later we had a holiday, and I brought three fishermen friends to the lake, having first made them promise that they would not take more than one fish each. Sure enough, within an hour and a half, we each had one fish and they ranged from a really fat six-pounder up to ten and a half pounds. As the area became known and frequently visited, the trout grew wiser and warier until they just wouldn't look at a lure anymore. The last time I fished there was with a German friend. We had acquired a flatfish that glowed in the dark after being exposed to artificial light, and we felt at last we had the ultimate weapon.

First catch at Lake Huascracocha.

We drove out to the lake and arrived about 8:00 p.m. It was a dark, freezing night in July with no moon, but about forty billion stars with a gentle wind that cut like a knife in the thin high altitude. We had brought only one rod and the new lure, so I took first go. My third cast I was firmly hooked on the bottom. Guns and I looked at each other and I could tell what he was thinking, so I gave him the rod to hold, and with a light in one hand, and the line in the other, I started wading into the lake. If the water had been two degrees cooler, it would have been solid. At last with little waves gently lapping just below my belt line, I got the lure free and made my way back to the shore. As I emerged from the water and the icy wind started blowing through my wet pants, my teeth started sounding off like castanets, and I told Guns that I would fish for twenty more minutes, then give him half an hour before going home, as even the hot coffee and rum didn't seem to help much.

My twenty minutes passed with no action, so I handed the rod to Guns. On his second cast he hooked the bottom and couldn't get the lure free. Finally, he ceased his efforts, and looked at my shivering form, and with German logic, pointed out that I was already wet.

Once more I found myself cautiously wading out into the black freezing water, with the line in one hand and a light in the other, my teeth chattering away like machine guns. I had just about got to the same depth and was on the point of freeing the deadly flatfish when the water exploded three feet from my face, and a ten-pound-plus trout jumped almost into my lap. I must have done a standing two-foot jump, I got such a fright I thought my hair would turn white. My inadvertent jump, though, had freed the lure, and I shakily headed back to the shore, my nerves still fluttering. I then told Guns if he hooked the bottom again he could damn well go into the lake himself, and that I would be quite willing to help him in if he dragged his feet. With this parting shot I sat down, sullenly shivering on the bank, sucking at the thermos of coffee and rum.

Guns didn't catch anything either so we headed back to camp empty-handed. The trip back was more cheerful as the rum was taking effect. I couldn't feel much below the waist, but the heater helped, and I sang flat for most of the way home, much to Gun's disgust as he appreciated good music.

Afterword

At 10:00 a.m. on the first of January, 1974, at the General Office in La Oroya, Harry Allen, Cerro de Pasco's Canadian Vice President and Manager of Operations, signed the decree transferring the C. de P. Corporation to a new Peruvian government entity called Centromin. The Peruvian Minister of Mines also signed the document, and then asked for the key to the company to symbolize the transfer. As no one had anticipated this, Mr. Allen' secretary, Sr. Elijade, quickly came to the rescue with his house key.

A few top level employees, including Mr. Allen, remained that year as advisors to the new bosses. Not long after that the last of the gringos came down the hill.

Mari-Jo Woolfe

Sterling silver hard hat and sterling silver miner's lamp with inscriptions to Brian from his co-workers.